"Building Excellence: Navigating Challenges in ACT/BIM Adoption for Major Project and Construction Quality Management"

Table of Contents

Table Of Figure

Preface

In embarking on this journey of exploring and elucidating the intricacies of Advanced Construction Technologies (ACT) and Building Information Modeling (BIM) for Major Project and Construction Quality Management, I am humbled and excited to share this knowledge with readers, practitioners, and enthusiasts alike.

This endeavor is a culmination of my academic pursuits, professional experiences, and a deep-rooted passion for advancing the standards of construction excellence. As a graduate in earthquake engineering from Jamia Millia Islamia, New Delhi, under the guidance of **Professor S.M Abbas,** my commitment to the field has driven me to navigate the challenges posed by ACT and BIM, offering practical insights and solutions.

Throughout this book, I aim to provide a comprehensive guide, not only shedding light on the technical aspects but also addressing the real-world challenges faced in the implementation of these technologies. I hope the readers find this resource valuable and transformative in their pursuit of excellence in construction management.

About the Author:

Er. Saiful Hasan, M.Tech (Earthquake Engineering)

Er. Saiful Hasan is a dynamic entrepreneur and expert in the field of earthquake engineering, leveraging his technical acumen to drive innovations in the construction industry. As the founder and leader of "The Contentpedia," a forward-thinking company established in 2017, Hasan has demonstrated visionary leadership in navigating the ever-evolving landscape of content creation, engineering solutions, and knowledge dissemination.

Er. Saiful Hasan, an alum of Jamia Millia Islamia, holds a Master's in Earthquake Engineering. His expertise extends beyond academic achievements, reflecting a fervent dedication to elevating the construction landscape. With a focus on ACT and BIM, Hasan's insights aim to empower professionals and enthusiasts to embrace innovation in construction.

Acknowledgments

In the creation of this book, I owe a debt of gratitude to my wife, **Dr. Shaheen Khan**, whose unwavering support and encouragement have been a constant source of inspiration. Her resilience and understanding have been instrumental in the completion of this project.

I extend heartfelt appreciation to my parents for instilling in me a passion for learning and perseverance. Their guidance has been the bedrock upon which my academic and professional journey has flourished.

Dedication

To my wife, **Dr. Shaheen Khan**, and my parents **Mr. faizul Hasan** and **Mrs Fareeda**, who have been my pillars of strength and inspiration. My **Professor S.M Abbas** has given me timely support whenever I ask for anything from basic engineering to modern time research. Your love and support are the driving forces behind my pursuit of excellence.

Chapter 1: Introduction

1.1 Introduction

Construction is an extremely low-innovation sector when compared to other industries. Building Information Modelling, or BIM is a new technology that has been brought to the construction sector along technological advancements. BIM is also known as Building Information Modeling (BIM), is a model-based 3D method of

managing and constructing construction projects. The goal of this bachelor thesis is to examine the ways in which BIM could be used to the

Figure 1: BIM is going to make everything revolutionized

most complicated projects of our moment, its impact on the management of projects and methods to maximize the benefits of this technology. This thesis is focused on large construction projects and the ways in which BIM is able to be integrated with ACT.

Building Information Modelling has been evaluated for its effectiveness and its inefficiencies for Construction Infrastructure Projects. This research was conducted to provide an authoritative and comprehensive study through both academic and industrial reports. Interviews with experts from industry have also been conducted to give additional evidence to support the implementation and implementation stages. The result was the creation of a strategy to combat the shortcomings and issues raised through BIM.

The research is all about the sociotechnical approach that not only considered the technology, but also the social and cultural environment as a context. The action-oriented quantitative and qualitative research provides "learning by do" through discovery, comparison and experimentation. The thesis describes a BIM adoption process adopted by SMEs in the project case. The thesis will contribute to the development of an adoption framework for BIM.

The complexity that is BIM, (acronyms and competing standards, technologies) can also lead to an inability to communicate, since individuals lack an objective understanding of the goals and the benefits. The research concludes by a synopsis and analysis of the findings and an approach to solution that is focused upon applying an objective approach on the needs and procedures. This framework also assists in the achievement of goals and will allow for agreeable measures to measure the benefits of BIM.

BIM (Building Information Modelling) is the product of increased collaboration between researchers, software developers, and practitioners of industry in recent decades. The architect, engineering and building (AEC), industry has now recognized BIM as a new technological shift. BIM has a significant impact on the architect, engineering and construction (AEC) industry.

It is also seen as an opportunity to help develop new thinking patterns. BIM has transformed the design and construction industry, resulting in more efficient and cost-effective projects. BIM represents the physical and functional attributes of an entire facility digitally. Estimators, fabricators and contractors can use this information digitally to control costs and reduce construction timelines, estimate waste, eliminate redundancy and minimize cost. It can help project managers at all levels in the project management process. The use of successful concepts is essential throughout all phases of a building's life cycle to ensure maximum benefits and comfort to occupants, without having to compromise the project's function.

BIM has received considerable attention by researchers in the developing world, but there have been few studies on how BIM can affect building projects. This research aims to eliminate any obstacles that might prevent BIM from being used in construction projects and find out how BIM was adopted in two construction projects. A comprehensive literature review was conducted to identify obstacles that may hinder BIM adoption. Since the 1970s, AEC has lagged behind in productivity growth compared to other production industries. This has been attributed to a mix of collaborative requirements in completing construction projects and the fragmented AEC industry. BIM has been proposed as a solution to these problems and improve productivity of construction projects. BIM adoption has been slow, and many obstacles have been identified that prevent widespread adoption. However, there is no one barrier which can be overcome individually to allow for more widespread BIM adoption.

The barriers are preventing BIM from being adopted effectively in many ways. Comparing BIM to 2D CAD, it is more efficient in handling the information related to a project or building. BIM adoption allows for changes to work processes which can improve the efficiency of construction projects. BIM adoption is more than just a technological change; it also requires a substantial shift in the work process to improve productivity. BIM can be used to achieve certain goals by improving processes. It is not an end in itself. The proper planning and quality management for construction has been defined in this process. The methodology has all the sections which has been mentioned to define the processing of BIM such as data collection and data analyzation.

Real estate and construction is one of world's largest industries, but it also has the highest fragmentation. This industry has a multi-disciplinary team working on a single project that faces many coordination issues. Information and Communication Technology (ICT), as an instrument to address these issues, has been proposed to increase the historically low productivity of this industry. BIM is a tool which allows the storage and re-use of domain and information knowledge during the entire lifecycle of a project. BIM's main function is to coordinate and integrate the sharing of knowledge and information between disciplines within a project. BIM can improve product quality and enable more sustainable building designs. Howell and Batchel stated that despite the widespread acceptance of BIM's economic and environmental advantages, the technology is still slow to be adopted. The purpose of this thesis is to understand the obstacles limiting BIM implementation in the construction industry. This will be done by studying the expectations of different project participants regarding BIM. (NBS, 2020)

Figure 2Best interior designing

Construction industry reimagines its self using innovative government tools. Construction is a key social characteristic that defines the quality and comfort of life for any nation's citizens.

The building industry has grown and changed dramatically in developing countries to accommodate local goals, as well as the needs of residents. The government prioritized the creation of affordable housing through a number of affordable housing regulations. By 2030, it is expected that over 250 large-scale projects will be completed in countries with low and medium incomes. In these countries, however, construction projects are often plagued by various problems (such as a lack of modern transportation and communication infrastructure or a shortage of required products). Tah and Carr claim that the construction industry has problems, which results in poor outcomes for developing nations.

A number of initiatives have also been put on hold due to a limited scale of investment. As a group, the construction industry of developing countries is not meeting the needs of governments, their clients and the society. It lags behind the other industries of those countries. The literature stresses the importance of "overall project success". Wolstenholme and colleagues claim that quality construction is essential for transforming the industry. Building information modeling (BIM), therefore, can be used to achieve success in the design and preliminary phases of construction. BIM is used to design and construct the built environment worldwide. (NIOH, 2020)

BIM, or Building Information Modeling, is "an intelligent 3D-model-based process" that provides architects, engineers, and construction professionals with the insights and tools they need to plan, design and construct buildings more efficiently, as well as manage them. The latent potential of BIM is to improve efficiency and effectiveness throughout the entire life cycle of a building. BIM is undergoing a radical transformation as stakeholders demand that technology be used to address recurring issues such as cost and time management, productivity, etc. BIM is now recognized as a lifecycle technology with a positive impact over the lifespan of any building. BIM has many benefits, but its full potential is not explored. Many studies have tried to unravel the Gordian Knot of obstacles to BIM adoption. They looked at factors such as acceptance, nature of hurdles and motivations.

Construction, particularly in less developed nations, is lacking systematic efforts to study the implementation challenges of BIM. Few studies examined the barriers to BIM implementation in industrialized nations. We asked "What are some of the biggest barriers in implementing BIM for low-income countries?" No previous study has tried to catalogue and rank these challenges. This is the first investigation to do so. The study could help stakeholders reduce waste, improve the quality of construction projects and increase the efficiency of the project by implementing BIM.

Aim for the research

The aim of the research is to show that how BIM is to facilitate and coordinate the exchange of knowledge and data between different disciplines in the project. It describes the actual condition of BIM and improvement challenges of BIM technology for the construction management.

1.3 Objective for the research

- To show the major elements to decide the exact BIM approach.
- The involvement and application of people with the various parties and BIM process.

- Showing the difficulties during the implementation and acceptance of BIM.
- The actual modelling of BIM will be represented as per the current demand of the construction project management.
- The last objective is to identified the biggest challenge to use BIM for the construction management.

1.3.1 Developing countries adopting BIM

BIM has become increasingly popular with construction experts around the globe. According to the National Building Specification, the United Kingdom (UK), Canada and Finland are among the advanced BIM countries. Building information modeling (BIM), both in terms of awareness and usage, has increased dramatically, going from about 10% in 2011 up to around 70% in 2019. McGraw-Hill reports that 64 percent of Australian businesses use building information modeling. Rodgers, et. al. estimate that 48% of SME's adopt BIM. The current BIM awareness is negative and skewed, with a focus on disadvantages, rather than benefits. (ADB, 2022)

A lack of management in building literature, expertise, or information can have a negative impact on the lifecycle of a project. BIM's graphical complexity and ease of use offer many opportunities for improving building performance. This tool can be used to manage the whole construction process. It is a useful framework for creating product models that are dense in information. These models are used to assess the efficiency of a building by taking into account its geometrical and thermal properties.

Cheng et. al. claim that building information modeling can improve the efficiency of MEP system maintenance. BIM can be used to sort out data such as the building type and geometry, which allows for better decision making. BIM can also be defined as a tool that is used to create a perfect model of the project's "as built" situation. Saka and Chan claim that the reputation of construction professionals as being slow in adopting new digital technologies, such as BIM, has hindered development and innovation. BIM can provide construction professionals with many advantages, including more efficient project management and execution.

BIM is a way to develop, merge, and maintain such databases that include vital data about a building or portfolio of buildings for operations and maintenance. Nieto-Julian and colleagues claim that BIM can help members of interdisciplinar cultural teams by making information exchange easier. Stransky & Dlask have shown that BIM can improve project performance, and help with decision making throughout the project execution. We find that BIM is a great way to strengthen the bonds between participants in a project. This was also emphasized by Eastman et al. BIM's value in cost management and estimation has been proven by further research. According to Chahrour and colleagues, BIM can save money by identifying major conflicts in the design phase prior to project implementation. Others have praised it for its intelligent automation of contract and collaborative nature. BIM is also important in the promotion of

20

sustainable building. The "Green-BIM" term, which aims to reduce the environmental impact from construction activities, is a good example. (Agenda, 2019)

Amarasinghe and Soorige assessed the use of BIM in Lifecycle Assessments (LCAs) and offered suggestions on how to improve BIM-LCA assessments. BIM's visualization features are one of its main selling points, allowing clients to visualize their project even before it begins construction. BIM allows the team of designers to change individual building aspects based on customer input. The visual interface provided by BIM is now seen as an essential tool for designing buildings, both during the initial design phase and during optimization. Lin and Hsu also used BIM for issue management and conception by using a web API. This shows how BIM can help visualize problems and the progress of a project. Raouf and colleagues claim that BIM impacts the project's lifecycle in a different way than traditional engineering project management. The project lifecycle is divided into phases, each of which has different professionals contributing at various times.

1.4 Problem statement

AEC has historically had a low productivity increase compared to other industries. BIM was introduced as a solution to this issue by many, however the rate of adoption is slow. Many researchers are developing theories to explain why this is the case. They focus on the different barriers that limit the use of BIM and the adoption of BIM for construction projects. It is not well-documented how these theories interrelate. It is important to understand how these obstacles combine, and what can be done about them. It is not clear who, if anyone, should be driving the adoption and development of BIM to combat this low rate of productivity.

1.5 Research ethics

Research ethics is very important. It is important to keep in mind that the research cannot harm or embrace anyone. All interested parties will be able to access the published thesis, which includes all of its contents. In order to address this problem, I made sure the interviewees were aware of my research goals and what I wanted to do with their contributions. Moreover, the research subject does not have a high level of sensitivity. Therefore, the contributions to my project are unlikely to cause any embarrassment or harm to anyone. I decided that all names would be kept anonymous and they will only be listed as the professional role of each individual in my projects. Interviewees aren't of importance to research, but their roles within the projects.

Figure 3: Advance construction technology is leading construction at different level

1.6 Hypothesis

This research problem may be used to formulate the following hypothesis:

Adopting ACT and BIM in the design management of major projects will increase efficiency, reduce costs and improve project outcomes. This, in turn, will enhance overall project success. The integration of ACT with BIM can positively influence design management in large projects, resulting to improved performance and project success. This research will investigate, analyse, and prove this hypothesis through the study of real-world projects that apply ACT and BIM. In recent years the construction industry has seen significant advances, driven by technology innovations that aim to improve project delivery.

These two innovations - Building Information Modeling and Advanced Construction Technology - have become transformative for the design management of major projects. Adopting ACT and BIM will increase efficiency and reduce costs. It can also lead to better project outcomes. The implications and benefits of integrating ACT and BIM into design management for large projects are explored in this long note. Adoption of Advanced Construction Technology (ACT) and Building Information Modeling for design management in major projects is a multifaceted way to improve efficiency, reduce costs and enhance project outcomes.

The technologies offer a comprehensive solution for the project's entire lifecycle, including design, construction and facility management. The initial cost of ACT and BIM training and tools may seem high, but the benefits of enhanced project efficiency and reduced costs over the long term make these technologies an important part of modern construction. It is vital that the construction industry continues to embrace ACT and BIM in order to remain competitive and grow.

1.7 PESTEL analysis

PESTLE is a framework that helps to evaluate the macro-environmental external factors which can have an impact on a project or industry. Let's perform a PESTLE Analysis to better understand all the factors involved in a study of ACT (Advanced Construction Technology), BIM (Building Information Modeling), and design management for major projects.

1. Politics Factors

- Regulations and Government decisions can have a significant impact on construction projects. Changes to building codes, environmental regulations, and safety standards may impact the adoption of ACT or BIM.
- **Grants and Funding:** The government may provide funding or incentives to projects that integrate advanced technologies, such as BIM. This could influence the adoption rate of these systems.

2. The Economic Factor:

- **Economic Condition:** A region's or country's overall economic condition can have an impact on construction projects. Budget constraints may result from economic downturns, whereas growth in the economy can increase budgets for projects and promote investment in new technologies.

- **Cost implications:** Implementing ACT or BIM can have a significant impact on the cost. Exchange rates and inflation are two economic factors that can affect project budgets.

3. The Social Factors

- **Levels of Workforce Skills:** It is essential to have a workforce that can operate and maintain ACT systems and BIM. The social attitudes towards training and upgrading can affect the adoption of new technology.
- **Technology Acceptance:** The rate at which advanced construction technologies are adopted by the public and industry can be affected by their acceptance.

4. Technology Factors

- **ACT & BIM:** Technological advancements and availability of the latest ACT & BIM hardware and software can have a significant impact on the effectiveness of the design management process.
- **Capabilities for Integration:** Compatibility of ACT or BIM with current tools and software. This is a technical consideration.

5. Environmental Factors:

- **Sustainability** The adoption of BIM can assist in the sustainable design and construction.
- **Resource availability:** Environmental variables such as the available of resources and construction materials can affect project feasibility.

6. The Legal Aspect:

- **Intellectual Property Rights** Legal Issues related to intellectual properties and data ownership can impact BIM adoption.
- **Insurance and Liability:** It is important to consider the legal framework surrounding liability for technology-related accidents or errors.

7. Ethics:

- **Construction Ethics:** The use of ethical construction practices and responsible sourcing can have a significant impact on project design.
- **Data Security:** When implementing BIM, it is important to consider ethical considerations regarding data security and privacy.

The PESTLE Analysis of a Practice Study on Adoption of ACT and BIM in Design Management for Major Project Delivery reveals a variety of factors which can affect the adoption of these technologies. Politics, economy, social factors, technology, the environment, law, and ethics all influence how construction companies implement advanced technologies. It is important to recognize and address these factors in order for ACT and BIM to be effectively integrated into design management. This will lead ultimately, and more efficiently and successfully deliver projects.

Integration of Advanced Construction Technology and Building Information Modeling into the design management of major projects can enhance efficiency, reduce costs and improve project outcomes. Although the benefits of this technology are obvious, it is important to conduct a thorough analysis in order to fully understand all the issues, challenges and concerns associated with its adoption.

1. The cost of implementation:

The initial cost of implementing ACT and BIM is a major challenge. The cost for software, hardware and training is also included. This can be a barrier for smaller firms or projects that have limited budgets.

2. Resisting Change:

Construction is a conservative industry that has resisted change for a long time. Both management and employees can be resistant to the introduction of new technologies such as BIM. The resistance to these technologies can lead to a slow adoption.

3. Skills Gap:

A skilled workforce is essential for the application of ACT or BIM. It can take a long time and be expensive to train existing staff in the latest technologies or hire new experts. In some areas, the availability of skilled professionals can be restricted.

4. The Interoperability of Issues:

It can be difficult to integrate ACT with BIM. It can be difficult to ensure that the systems are integrated seamlessly with each other and existing software. Data transfer or miscommunication issues can cause errors, and delay projects.

5. Secure Data and Privacy:

BIM requires the management and collection of vast amounts of data. It is vital to ensure the privacy and security of these data. Loss of project data, or breaches of confidential information about the projects themselves can lead to serious reputational and legal consequences.

6. Learn Curve:

These advanced technologies have a steep learning curve. The project teams will need some time to become familiar with ACT, BIM and adapt. Productivity may be affected during this phase of learning.

7. The Project Size and Complexity:

The scale and complexity may affect the suitability of ACT or BIM. These technologies are well-suited to large, complex projects. However, they may not be as useful for small projects.

8. Regulation Compliance:

Construction industry regulations and standards are numerous. Implementing ACT or BIM requires compliance with new rules, which can make it difficult to navigate the regulatory landscape.

9. ROI:

It can be difficult to evaluate the ROI of ACT or BIM. It can be difficult to measure the impact of cost and efficiency reductions on long-term project outcomes.

10. The Industry Adoption rate

Globally, the pace at which technology is adopted in construction varies. There may be a shortage of support and expertise in regions that have not adopted ACT or BIM.

Adoption of ACT for major projects and BIM in design management offers a variety of benefits, ranging from improved efficiency to better project outcomes. The critical analysis shows that this adoption is not without its challenges. To overcome these obstacles, a strategy is needed that takes into account budget restrictions, the resistance of workers to change, skill-development, data security and industry regulation compliance. Implementing ACT and BIM successfully depends ultimately on a carefully-planned strategy, and a willingness to overcome the inherent obstacles in the integration advanced technologies in the construction industry.

Construction is a vital part of Sweden's development, growth and prosperity. The construction industry employed 300.000 people in Sweden, which was 6,7% of the total Swedish workforce. The construction industry in Sweden employs 305 000 people, which is 6,7 per cent of the total Swedish workforce. Investments in this industry totaled 266 billion Swedish Kronor (43.2 billion AUD), or 8 percent of GDP

The UK Government prescribed a UK Building Information Modelling Mandate in 2011 as part of its Government Construction Strategy. (HM Government 2012). UK Government's intention was to bring together project stakeholders and reinforce them at project delivery with the requirement of "collaborative BIM" for all centrally procured construction contracts from 2016. This mandate also sought to save '20%' by adopting and advancing digital methods, and

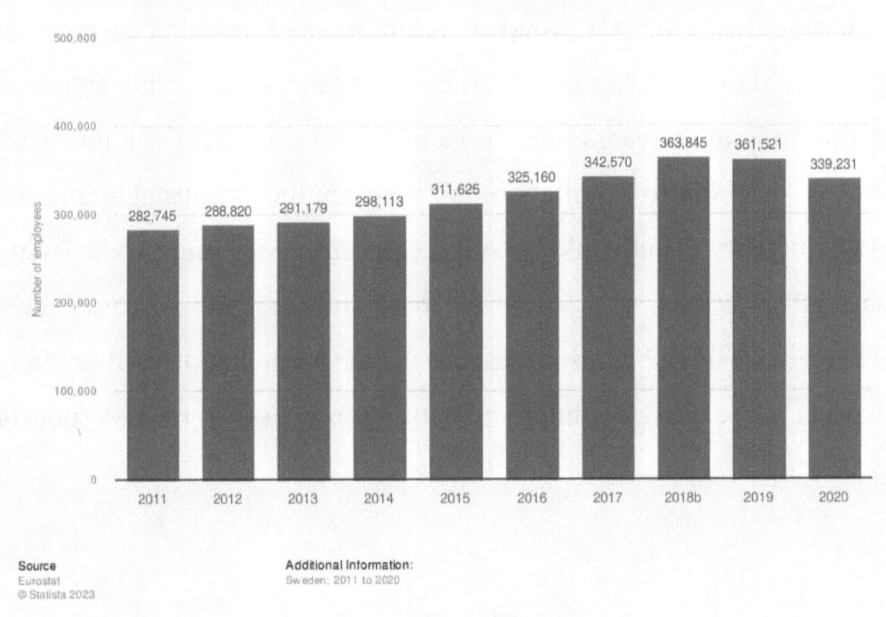

Figure 4: Construction industry employees in Sweden in 2020 (Source: Statista)

achieving efficiencies through adoption. The UK BIM Framework Roadmap will also include new digital working methods. It was expected that, over time, the digitisation of projects, starting with initial feasibility studies and continuing through execution, delivery and completion stages, would lead to efficiencies, along with the advancements in BIM and Digital Engineering.

Rapid development in information and communications technology has led to the creation of some promising new digital tools. Recent years, enabling technologies like building information modeling have appeared. BIM refers to computer-assisted, product-oriented three-dimensional design processes and technologies in the AEC industry. BIM allows a group of actors to work together in order to create a 3D model of the construction project. BIM in this article is defined as both a technology and a broad process. BIM is a powerful tool that can improve collaboration between the various actors in the AEC industry. This is likely to increase efficiency and productivity, as well as reduce cost. Although both industry and research have invested considerable resources into the development of BIM in many aspects, the basic implementation process remains poorly understood.

The implementation of BIM, for example, involves a lot of organizational changes and can be a challenging process. The construction industry has its own unique characteristics that set it apart from the rest of the industrial world. The characteristics of an organization are important for the way it changes. Construction, for example, is a diverse industry with many different companies and professionals. The construction industry is often divided into separate projects with a few standard tasks and seldom the same personnel. Often, people from various firms will work together on a specific project but not again. The complexity of the project will likely have a significant impact on how BIM implementation is done. Harty explains that the complex construction context is characterised by an inter-organizational collaborative approach, power distribution amongst cooperating organizations and a project-based method.

It follows that BIM implementation at the inter-organizational scale is beyond the control and influence of any single party who could ensure an uniform implementation. In previous research, implementation was primarily studied at the industry level or within a company. Fewer studies have examined implementation on a project-level. This refers to the "what happens" in a BIM implementation into an organization of project firms. This paper aims to identify the different forces that drive and restrain BIM implementation. This will lead to the following research question: Which factors are important for BIM at project level? We will not be examining the consequences or effects of the implementation but instead the process itself. Implementation is the process of putting BIM to use.

Investigated is the Norwegian Directorate of Public Construction and Property. Statsbygg is responsible for providing public enterprises with functional and appropriate premises. Statsbygg is responsible for providing guidance on the acquisition and lease of premises, and acting as building commissioner in relation to new buildings. Statsbygg was the building commissioner for a renovation of a government building. In this case Statsbygg requested the use of BIM including some Lean-construction-inspired principles in the design phase. Statsbygg played a key role in this case study which will later be detailed. Statsbygg was able to formulate how BIM would be implemented in the project. Statsbygg's desire to drive the Norwegian Industry in BIM and other areas is probably the reason for this.

1.8 BIM for large construction

Statsbygg is responsible for a wide range of tasks, including acting as building commissioner. Statsbygg is a pioneer in the Norwegian Construction Industry for the implementation of BIM. Statsbygg published its first BIM guide in 2008. The same year, they called for the implementation of BIM as part of a pilot in Western Norway. Statsbygg is a pioneer in the Norwegian construction industry in terms of BIM implementation and development. Statsbygg also participates as an industry partner in a Norwegian ongoing research project named SamBIM, which is funded by the Norwegian Research Council. The research project was developed by a collaboration between Norwegian research and industry partners. This project aims to test new technologies and organizational forms in the construction sector, such as Building Information Models.

This project aims to create value and innovate in society, AEC industry and companies by improving BIM supported processes and collaboration on real projects. Statsbygg contributed to the research project by submitting a case study. Statsbygg chose this project because the project would start on time and project management was interested in participating. We will examine in the next sections how Statsbygg implemented BIM, including Lean construction methods. This case is about the refurbishment of an Oslo public building. It was built in 1974, and is largely unchanged. Construction of 1700m2 was required, as well as rehabilitation of 4,900m2.

Figure 5: BIM for large construction

Statsbygg decided that the project would be part of SamBIM's research project even before the design phase. Statsbygg decided to have higher BIM goals than usual for the design stage (the preliminary and schematic part according to Statsbygg's projection model). After the project was out to tender, some of the project members from Statsbygg suggested that as a part of the elevated BIM ambitions, it would be interesting to try out some Lean-construction-inspired principles in the design phase as a part of the BIM-plan. The tender documents, which were distributed weeks before the project was awarded, did not mention this.

The idea was essentially to test out the co-locations of the design teams a few days a week in the same office, along with some "lean" working methods. Statsbygg did not discuss or specify the latter at the time. Statsbygg was new to the idea of co-locating a design team and introducing some innovative working methods. Statsbygg had little

37

experience with such a working method, yet they were interested in gaining more knowledge from the perspective of a building commission. They wanted to know if such methods of working could result in a faster and better design process.

A design team has been selected based on an open tender. A design team consisting of small and medium-sized Norwegian companies based in Oslo was selected. Statsbygg explained in interviews that the group chosen was primarily based on its extensive BIM expertise. In addition to the proposed solutions and team experience, cost was also a factor in making the final decision. Statsbygg organized several meetings after the contract had been signed and the project started. These discussions covered issues related to the project and BIM in particular. These meetings were organized by Statsbygg's project manager, assistant project manager as well as the so-called Change Agent. Statsbygg's change agent, a Statsbygg staff member with a specific responsibility for SamBIM activities within the company.

1.9 New technology with a case study

Statsbygg presented the idea to the team during one of their first meetings. They wanted to test the co-location in the same office of all the designers as a part of BIMplan. Statsbygg made the proposal in a very open-ended manner. Statsbygg didn't come up with a set of options for how to do this. Statsbygg welcomed open, enlightened discussions with the design team about how to co-locate and set up the work plan. After a few meetings, Statsbygg agreed with the design team to use the BIM plan below for the design phase. Statsbygg representatives and the design team were going to co-locate one or two days a week at an office near one of the architects' offices. Several of my sources told me that this form of work was inspired by Lean construction and the principles of so-called Virtual Design and Construction. VDC is a term used by Stanford University's CIFE Center to describe a Lean-influenced concept. There are several overlaps and similarities between Lean construction and VDC. One of the most important similarities is that both methods focus on activities which add value to a project, minimize waste and focus on pulling mechanisms.

Under the umbrella of "VDC", a variety of techniques and methods have been developed. In this instance, two Lean-constructions methods or tools originating from VDC were planned to be applied. These are Integrated Concurrent Engineering and Big Room-Organization. ICE is a series of parallel, co-located design sessions with the central goal being better collaboration. In so-called ICE session, the relevant parties are gathered together in a large room, where they use computers, databases, and SMARTboards simultaneously. The big room in this instance was to be equipped with desks around the SMARTboards, in an open office of a sizeable proportion. It was also planned to equip two adjacent offices with separate meeting rooms, telephone calls and so on. Statsbygg agreed on a schedule for the weekly meetings as part of the BIM plan, which also included co-location. According to the weekly schedule, every Wednesday the design team should have all the disciplines present if they are asked or need them. On Thursdays the entire team should be there as well as Statsbygg representatives. On some Thursdays, the team planned to meet with representatives of the users. This form of work was intended to strengthen the inter-disciplinary collaboration, speed up design and reduce response delays. In order to test the tool, Statsbygg and the design team planned on using a "planning matrix".

It was intended that the matrix would serve as a calendar between co-located meeting. The matrix was a whiteboard arranged by dates, disciplines and activities. To assign tasks, Post-it Notes in various colors were used. It was planned that this tool would also be used to create what is commonly known as "actions items". During meetings, action items are generated when an important task is identified. It is important to document the action and assign it to a person, typically a group member. The group is then obliged to report the outcome of the assigned action. It was hoped by the team that the combination of using the action items and the planning matrix would lead to a successful planning and execution phase of BIM.

Chapter 2: background/Overview and challenges with previous study

2.1 Overview

Thompson and Miner describe the basic idea behind Building Information Modelling. If all data relevant to a particular project was stored online, then it could be executed first in a virtual world. The addition of dimensions such a s time and cost to the model allows for a quick analysis of costs versus

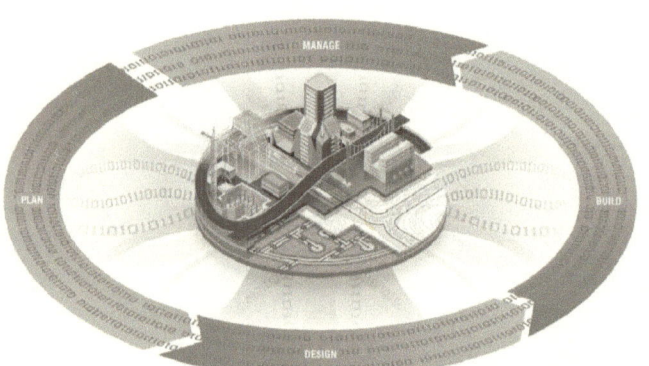

Figure 6: BIM implementation in an organisation

benefits. In the development of such models, project stakeholders can be involved earlier than is currently possible. They can bring their engineering and business knowledge to the planning and design phases, improving the coordination of all phases. This BIM model will be a parametric, data-rich representation of the building that can serve as an archive for all the data needed to suit the needs of different users throughout its entire life cycle.

BIM Implementation in an organisation requires costs such as purchasing the BIM software, and teaching personnel on using the software. BIM implementation can be hindered due to the initial costs associated with updating software, buying new equipment, and training employees. BIM cost of implementation was not more expensive than the benefits therefore BIM was seen as an expense. BIM was not able to provide any incentives in the form of financial rewards to promote the adoption of BIM. BIM cost is high due to of the expense of BIM experts. Laws were found to as the second largest obstacle for BIM implementation. BIM is not widely used due to the fact that there is no law which require it to be used at specific sizes. Work contracts were insecure and therefore BIM was no longer a factor. The law was applicable to all aspects that included BIM standards, as well as contract for projects. Countries that didn't have BIM standards for national BIM standard were the ones that had none BIM laws. Demand was the third obstacle in BIM adoption. BIM wasn't yet widely used by the construction industry. BIM wasn't widely utilized due to the ambiguity surrounding the immediate advantages, particularly in the planning stage. The perspective of BIM, which didn't reduce the time that was spent in drawing, compared to traditional drawing methods, is another reason for the reason BIM isn't utilized.

Refusal to adjust to changing circumstances is due to an underlying cultural character. Construction workers were hindered with a lack of enthusiasm when it came to establishing new processes and with training personnel. It was challenging to establish the new way of life in an environment that reflected the same value system. Workers' attitudes towards technology was that they believed to be difficult and complicated to apply in their area. Construction workers were resistant to changing since they had become comfortable using the old methods. Benefits of BIM were, according to them, weren't real and couldn't be felt directly. It is going to take a huge shift in the culture to alter the way people view BIM. Interoperability made the process more challenging. Some categories are not as frequently used than the ones mentioned in the preceding paragraphs, like the management of expertise, processes and knowledge, as well as standard skills, and education were deemed important to emphasize in the context of BIM obstacles to implementation.

BIM can be used to Indonesia to help overcome the obstacles. The obstacles could be classified into fifteen distinct types. First, it is necessary to develop guidelines that can serve to guide contractors as well as consultants. This is due to the fact the fact that Indonesia is a nation that is in the process of development, has control over most of the projects it develops. The government regulations could help in overcoming the obstacles. It is essential to ensure that the government shows an unwavering commitment to implement BIM. BIM adoption in Indonesia is easier through the creation of BIM education, BIM standards and certifications, BIM education and guidelines for groups of stakeholders. Lack of understanding and the unwillingness of people to experiment with new ideas is likely to change. It is becoming more crucial to have the ability to utilize the latest technology to enhance the flow of information.

Xu et al found that the presence of a local partner is crucial for the development of an international AEC firm. Based on this research the strategic partnership that include local construction firms or design companies are the most effective collaboration models. Local partners can provide their services for less and help the foreign business to overcome licensing challenges. An organized and clear method of communication for resolving conflicts and the ability of the team members to work together and define roles, the dedication to win-win relationships and constant monitoring were identified as crucial to partnering's   successful.

The design-build method of project execution as a method of getting into the Chinese construction sector is a possibility but the Chinese institutional environment is an obstacle to wide-spread acceptance (Ahola, 2021).Ling et al conducted a study to determine the most important factors that help AEC firms secure contracts with China. The capacity of the firm to understand the requirements and demands of its clients was identified as the most important aspect. This makes it challenging to expect AEC firms from overseas to be successful if they're not able to effectively communicate with their customers to satisfy the needs of their clients. Engagement and communication are the most important elements in determining the effectiveness   and   the   success   of   Chinese   AEC   firms.

In order to be able capture the requirements of their customers and understand their needs as the key element to their growth Foreign AEC firms must be aware of the different cultures. It is a fact that many do not consider. The study reveals that it is crucial to understand the Chinese culture, the language, Chinese workplace, characteristics of the construction industry and the dominant culture of organizational. FDFs are able to overcome the cultural barriers by establishing lasting relationships with Chinese businesses. While foreign AEC firms strive to better understand the cultural differences between them, they must prioritise building confidence. The trust factor is an essential factor in generating business benefits like lower expenses as well as shorter time frames and better performance. Trust and communication are essential for establishing partnerships with local Chinese institutions or companies.

The other risks that foreign AEC firms could be vulnerable to in China include clients' erratic behaviour, policy changes often, language issues contracts, transfer of technology, along with the influence of government and financial institutions (Pheng, 2019).The issue has been extensively discussed how important it is to have a well-organized construction design management system for the success of a project. Design management has been plagued by ineffective communication, poor documentation, inadequate or missing information, unbalanced resource allocation, inconsistent choices, and a lack of coordination between disciplines.

Poor design management can result in document errors and revisions, as well as projects that are overrun, reduced efficiency and higher costs. Design can be a major issue during the subsequent phases of construction. It affects management systems and the capacity to realize the full potential of a variety of initiatives. A better control of the design phase could result in improved construction ability and tangible improvement in terms of the timeframe, quality and cost in the event that designers implement construction ability strategies (Li, 2022).

Iterative design is the process of designers finding issues, sharing information and thoughts with one another and then implementing their ideas and resolving the issue. In order to improve the efficiency of design, it is essential to improve the design process, particularly in the initial stages of conceptualization or the initial phase. Design is usually seen as a continuous discussion between problems and the solutions (IDF, 2022) . Designing isn't an end-to-end process that is fixed since design issues are difficult to define and come with many possibilities for options. In reality the design process, as well as budgets are usually restricted. Therefore, it is essential to control the process of design (Bender-Salazar, 2023).

The design of the construction is a continuous process that has numerous interdependencies. This requires the participation of all participants as well as the coordination of all project participants. It's a fundamentally collaboration process, which relies on the sharing of ideas and information with others. Coordination and communication throughout this stage is therefore crucial. Flager and coworkers found that designers can be spending as much as 58% of their time during the design phase coordinating information and coordination. This involves manually integrating specific representations of the design as well as analytic models (TIDF, 2019). If an efficient management of information and coordination approach was adopted the designers could have the ability to devote longer performing analysis and design work, where projects have the highest worth (Akponeware, 2018).

Collaboration and communication when designing and carrying out a design can be seen in the variety of methods that focus on processes and are used to improve and enhance the management of construction design. Senescu et al. introduced the Design Process Communication Methodology in an effort to improve the effectiveness and efficiency of collaboration, communication and sharing. Choo et al introduced DePlan as a method for managing design which integrates design into the design phase.

DePlan incorporates Analytical Design Plan Technique to minimize rework in the design process and also it also incorporates the Last Planner System, a method of planning collaboratively that is commonly used in Lean Construction. Rosas implemented his Design Structure Matrix into the construction design in order to reduce uncertainties in design management. Hamzeh et al. presented an analysis of an example of design that utilizes The Last Planner System. The authors reported that despite the difficulties posed by the Last Planner System for designers as well as the change to a method of planning that this brought they were successful in implementing the new system. Process-Parameter-Interface (PPI) model is applied to address the design management issues of improved design process scheduling and efficient collaboration (Knotten, 2020).

Design Interface Management System(diMs) is a method that manages the iterative process of design for large-scale projects is currently being developed. The Dependency Structure Matrix (DSM) can be utilized to develop design plans and also to illustrate information exchange in the design phase is a tool to create the schedules. In order to improve cooperation, communication and communication within teams involved in designing and increase dedication to the Project, the collaborative Manage Method (CDM) is employed (Li, 2019).

Modern management of design in China is a relatively new concept. But, given China's rapid expansion of its technologically advanced society and its economic growth over the next few years it is likely that we will witness rapid developments. In China the local design firms are independent consulting firms that employ a variety of methods of design management which have been created ad in ad hoc. FDFs operating in China have to collaborate closely with their Chinese partners and other stakeholders in order in order to apply these innovative methods of design management. It could pose many issues. These challenges can be technological as well as legal and cultural (Ibn-Mohammed, 2021) .

Building Information Modelling, a process that is developing and able to efficiently store and manage the entire project's data (physical properties as well as functional attributes) in a single database. BIM tools are utilized for design purposes to carry out various functions such as visualizing design and generating new designs in a short time and predicting the performance of buildings, checking model integrity, and reporting, which allows the communication of designers and construction experts. BIM is becoming increasingly used in the construction industry. There are numerous documented advantages, such as a decrease in project costs and duration and enhancements to coordination, communication as well as the quality of. BIM implementation in the design phase can bring numerous benefits, such as enhanced coordination and communication across disciplines and a sharing of knowledge among the design team and other the other stakeholders. (Matusova, 2018)

BIM is, on the other side is a management of information tool that simulates the construction and design process instead of merely a visual representation. BIM surpasses 3D modeling and allows designers to choose the best design for their needs by modeling different options for design as well as their impact on the project. BIM facilitates communication regarding objectives as well as design modifications and issues. BIM is also a great tool to enhance the quality of design by finding conflicts among disciplines and then resolving these. It will also decrease coordination mistakes. Popov and co-workers demonstrated how BIM could be used to support a management system for designers that lets them efficiently share information, avoid errors in translation, miscommunication and prevent the loss of data. It results in an improved design process. BIM's ability to create quick design options and sketches as well as volume take-offs and provide an synchronized platform for various disciplines, in addition to the ability to capture more value aid in reducing non-value-adding design tasks within AEC. AEC sector (Zhang, 2020).

It is worth noting that the Chinese government, in contrast to other nations, is yet to make any regulations for national use that require BIM implementation. This means that BIM acceptance in China in the past decade is heavily influenced by the market. The BIM acceptance gap between AEC sectors that are "advanced" BIM users and the BIM industry who are in China may pose some issues for FDFs who operate or collaborate with Chinese AEC firms. The China Development Outline 2011-2015 stresses the necessity of speeding up the adoption of BIM as well as pushing ahead IT standards.

However, there is a consensus it is believed the fact that BIM in China is not mature enough to be implemented throughout its life period due to concerns about technology maturity, interoperability, cultural shifts in the industry, absence of a BIM standards, in addition to the need for education and training. Liu and Zhang discovered that even though BIM usage has grown in China in recent years, its use and use is still restricted. The authors analyzed 10 large projects and discovered that BIM is most often used in the stage of detail design. An investigation conducted in 2011 of China found that just 22% of the AEC experts surveyed were either familiar or extremely familiar with BIM. A lack of BIM understanding and the lack of commitment from management are two major factors behind the lower BIM acceptance rates (Herr, 2019).

Yung et al. reported the results of a BIM research study based on BIM of BIM-based design in China. Utilizing BIM will not necessarily reduce duration of design due to the fact that 2D designs remain heavily in. This is mainly because of the need to provide 2D documents to regulatory agencies to get approval. But, it's difficult to use BIM applications to create 2D drawings for shops that are in line with the industry standard in China. BIM is still able to lower the costs of electrical, mechanical as well as plumbing (MEP) as well as enhance the quality of designs through the reduction of change orders. Current Chinese laws can affect the use of BIM. Cao et al. investigated the 106 BIM projects across China.

The results revealed that BIM is used mostly to visualize buildings with the aim of displaying complex structures or to detect conflict between different building systems. The results of the study also underline how more collaborative/integrated project delivery systems, which are currently very limited, can support and be supported by a deeper BIM penetration in China. The research revealed that the most important BIM-related factors in China are the motivations of architects (economic advantages as well as efficiency and efficacy of BIM acceptance) as well as technical concerns (improvement in compatibility, and the integration of BIM with other widely accessible software) as well as BIM capabilities of the rest of the team members (Vass, 2019).

Eastman et al. stated that a Building Information Model was constructed by intelligent digital assembly with embedded knowledge about parametric attributes and characteristics. Kymmell has described BIM as an intelligent digital model that is connected to other project management software (e.g. Schedule, budget, or cost estimation) facilitates design optimization and constructability for all project stakeholders. Many authors support the process-oriented view of BIM, stating that BIM encompasses much more than a digital building representation technology (Jiang, 2021).

Succar describes BIM as an interconnected set of policies, processes and technologies that produce a new method to manage design, construction, and operation of construction projects throughout their entire lifecycle. Gu & London claim that BIM is a digital management approach in construction using IT. Aouad & Arayici stated more recently that BIM was the use of information and communication technologies to facilitate and streamline all processes involved in creating a safer, better environment. BIM is a process that uses technology to produce intelligent, information-rich models. This supports life cycle management of built environment projects. (Satyanaga, 2023)

According to an analysis of BIM, software companies that produce the software and technology for BIM implementation are in favor of the view "BIM is a technology". The "BIM as a process" or "activity" viewpoint is supported by the professional institutes of the industry, which is the main driver for the reengineering industry practices in order to support and incorporate BIM enabled work. Academics involved in BIM research, development and industry adoption describe BIM as "a system" and "a holistic approach". (NAIM, 2019) True BIM tools have an integrated database, where data represents the model of the project and reports are queries or views.

All plan views, sections views, elevations and callouts; as well as perspectives, are all live in a project model. Project teams and construction projects are divided into silos based on their discipline (architects, contractors, engineers), which operate independently and separately at every stage. When a project moves from the concept to the further development stage, information is moved "over the walls" because the CAD data cannot be integrated into a centralized point of truth, which limits collaboration between project participants. This practice also leads to an unlimited number of drawings, documents, versions, and change history which is difficult to maintain and manage separately. This leads to a lack of coordination and communication, as well as errors, duplications and process bottlenecks (NAIM, 2020).

The overall process of CAD is therefore unstructured and error-prone. It's also non-collaborative, inefficient, and not collaborative. BIM workflows, on the other hand, are built upon the idea of collaboration and integration to create a centralized repository of information for the entire project. BIM models enable project teams to visualise and solve design issues throughout the entire work process by using parallel and concurrent work streams. BIM models reduce the design errors and improve quality of multiple 2D documents. This leads to less changes in construction (Zabin, 2022) .

BIM can be used to produce schematics and design details, improve the presentation of a project to the client for better decisions and a more clear understanding of what work is to be done. Autodesk said that BIM was used to analyze energy efficiency and sustainability. It also helps with cost estimations, budget and schedule information. The cost and duration of construction can be affected by timely decisions and schedule changes during design. BIM can make these changes quickly, easily and accurately. This reduces the project's time and cost. Olofsson highlighted some of the benefits of BIM in Healthcare projects during their conceptual stage. These include quick visualisation, good decision support in project development processes; accurate automatic updating; reduction of man-hours for space programs; improved project team communication and an increased sense of confidence in scope of work (Pan, 2023).

58

Hardin, Eastman et al. agree that BIM can be used to analyze the operation of a building within a BIM-enabled environment. Anumba and colleagues state that BIM extends beyond the construction and design phases to facility management, with the goal of maintaining and maximising the value of the facility. Gu et al. stated that facility managers would need to keep an accurate space inventory (Lin, 2020). BIM offers a range of new software applications and tools that can be used for creating, analysing and managing BIM models at all stages of the project lifecycle. BIM tools store information in BIM models databases, which remains there until it is modified or removed. This allows information to evolve from a schematic design into a detailed one and beyond. Interpretability is another crucial feature of BIM software. It allows information to be exchanged without the use of any applications or tools. This is essential for effective team collaboration. BIM tools can exchange data in their proprietary formats or native formats (Kjartansdóttir, 2020).

BIM software is a complex tool that requires specialized developers and vendors. Users in the construction industry must rely on these companies to provide BIM applications and tools for varying uses. Software vendors have been driving development of BIM software and tools since the advent of CAD. These technologies allow for design flexibility and computer intelligence in creating effective information models. BIM-compliant software is available in hundreds and the number continues to grow as BIM becomes more popular. In the literature, many authors have divided BIM tools by category, including preliminary space planning, massing, sketching, preliminary environmental analyses, and preliminary cost estimates. Other categories include BIM design, structural design, BIM Construction, fabrication, environmental analysis, construction management, cost estimation, and specification tools. This table provides an overview of key BIM applications, tools and uses at different phases of a project lifecycle (Abioye, 2020).

BIM's benefits and productivity gains are widely acknowledged. These are becoming increasingly apparent as technology and process adoption matures. BIM implementation and adoption on actual projects are not without their challenges, which is slowing down the BIM take-up in the industry. BIM is a disruptive technology because it has a wide range of effects on working practices, contractual and legal settings, data ownership, security, and insurances. There are also issues relating to software and hardware, training, learning curves, and people's resistance to change. Many authors have addressed the BIM adoption and implementation challenges. These are mostly related to: (1) Organisational BIM adoption, (2) Project-process level BIM Implementation and (3) Technology related BIM Challenges (Salleh, 2023).

The challenges of BIM adoption and implementation at the organisational level are related to policy changes, changing working cultures, costs, and people. Fear of change is the biggest barrier for BIM implementation and adoption. When people succeed at something, it is a psychological phenomenon that they are comfortable doing the same thing again and again. BIM brings about multi-directional changes to the way people operate and work. Organisations are bound to encounter resistance against this change. There is also a lack of documentation on BIM workflows or standards, institutionalized quality control for BIM models, and BIM-specific mitigation or risk identification policies that are widely accepted and endorsed. Insurance companies and large organisations are still not able to write policies and price insurance for BIM project because historical data and empirical evidence of BIM productivity improvement isn't available (Sahil, 2020).

London et. al. highlighted that the adoption of BIM at the organisational level is determined by the relation between the current working practices and the future BIM scenarios as perceived in the organisation's policies. If an organization is at level zero BIM and wants to reach level two BIM in the next couple of years it will be difficult to meet this target. It would require extensive planning, commitment and resources to implement BIM. The key question is how BIM will change current working practices. The common belief is that existing processes, workflows and business models must be changed if an organisation wants to adopt BIM. This is incorrect, as BIM is flexible and can be tailored according to the company's needs. Some authors, however, have stated the opposite, saying that BIM adoption is dependent on significant changes to current business practices (Zaia, 2023).

BIM adoption is hampered by information bottlenecks and lack of content in project vendor products. There are also challenges with BIM application and understanding. Froese also stated that BIM's impact on project management processes and practices cannot be achieved without changes to the organisations and skills of project participants. Henrik & Linderoth emphasized the necessity of redefining the project roles within the current construction network from the top management down to the bottom supply chain in order to meet the BIM management process requirements (Kineber, 2023).

The final challenge is the implementation of BIM at product level. These challenges are technical in nature and relate to BIM development and its application within projects. BIM is an entirely new technology that has triggered the implementation of BIM in the construction industry. Many companies see it as a simple combination of software applications. Shafiq et al. state that the BIM technology is maturing and getting more suitable for its purpose. Researchers have stated that BIM is available in many places but the main challenge is the widespread use of BIM. The technology barriers are not just software or hardware, but also the acquisition, training and learning of new technologies to help implement BIM, like web portals and geographic information systems. Construction researchers are aware of the need to integrate BIM data and software throughout the entire lifecycle of BIM projects (Shojaei, 2021).

In the construction sector, productivity and production are terms that are used often interchangeably. However, there are significant differences between them. Production is defined as work that is done to achieve a goal or task, whereas productivity is a measure of average efficiency that is expressed by the ratio between inputs and out-puts in achieving the process. In the construction sector, productivity is called productivity rate (i.e., unite output per unit input). Low productivity in a production process means that resources are wasted (time, materials, efforts etc.) and it is an inefficient process. High productivity means that the resources used are efficiently utilized to get maximum results from given inputs. Simply put, higher productivity is getting more done in the shortest time and at lowest cost   (Kunkatla,   2022).

In the construction industry, productivity is of great importance. It is the best indicator for how much money and time will be spent per task and on the final project. Chelson quoted Warren as saying that "productivity in construction means completing construction work with unit rates less expensive than other than published estimates handbooks and better than estimated for the project". Construction productivity is defined by the U.S. Department of Commerce as dollars of output per hour of labor input. There are many types of inputs that need to be taken into consideration in construction, such as land, material, machinery, tool, and human resources. The Construction Industry Institute reports that the US construction sector spent $600 billion, or 57%, of its resources in 2008 on activities with no   added   value   (PiC,   2020).

In a typical construction project, the architect is responsible for designing and producing construction documents while engineers take care of technical details. This happens in separate disciplines. Lead architects are often responsible for the coordination of design. They perform regular coordinate checking using layers within CAD to eliminate any design conflicts. This approach, however, is inefficient, slow and ineffective because the information about the project cannot be integrated by using the traditional CAD file in sequential stages. This leads to many design conflicts that are realized during construction, leading to time delays, additional costs, and rework. The situation is drastically improved if design and construction are overlapping, which can be achieved by BIM-enabled projects or integrated project delivery (RIBA, 2020).

Contractors have traditionally used various project management tools, including WBS (Work Breakdown Structure), to plan construction activities. WBS (Work Break Down Structure) is a hierarchical, deliverable-oriented decomposition that breaks down the tasks to be performed by the team in order to achieve the project goals. This can hinder effective coordination across the entire project by breaking the work up into smaller packages without evaluating the relationships between them. The most popular scheduling software, including Primavera P-3 and P6, MS Project, uses a variety of techniques to produce optimised schedules. Schedule accuracy is important to not only determine the project's final completion date but also control the relationship between subcontractors and suppliers on the site. Unreliable schedules can lead to delays, material waste, wasted time, and labour waste (Wrike, 2020).

Contractors also have a responsibility to assess the constructability of designs and determine how they can be built using the most effective methods. Chelson reported that improving constructability tends to reduce the cost of construction by 6%-10%. Latham stated that designers, contractors, and engineers could collaborate to review constructability, saving up to 20 percent of construction costs. This is a significant improvement in productivity. A project that undergoes an ineffective constructability assessment and poor scheduling could result in a 9% decrease of performance for each task disrupted. A ductwork area is built and it's later discovered that the pipes don't fit. It is especially evident in the case of mechanical contractors, who suffer from the low productivity on site due to poor scheduling and design conflicts (Putu, 2019).

A good document management system is another important factor in contractor productivity. It allows the latest and correct information to be sent to all stakeholders at any given time. These systems include Meridian, Primavera and Bentley, as well as project websites like Asite and 4Projects. Delays or incorrect information can lead to disasters, and reduce productivity. Contractors are faced with several non-value added activities in traditional practices due to sequential procurement and negative contractual relationships. This includes waiting for approvals and submittals. It also involves requests for information, design changes, and other non-value adding paper management tasks. Material procurement and site management are also a productivity issue for contractors. These activities cause waste on the job and reduce productivity. In Europe, it is believed that 10% of the material on-site is thrown away. This figure is even higher in countries like KSA (NAIM, 2020).

The lean construction technique emerged to focus on activities that add value and reduce or eliminate unproductive work. Lean construction principles were borrowed by the automobile industry. Toyota Production System is a system that promotes continuous and concurrent improvement in the construction process with minimal cost. Lean construction basically aims to decrease the amount of non-value adding activities such as approvals waiting times, material waiting times, etc. that do not directly contribute to productivity. Lean construction has been shown to improve productivity by reducing labour hours, and wasting less time. This is a technique of management that relies on strategic partnerships and collaborations between construction stakeholders (Demirkesen, 2021).

To gain an advantage, construction executives and project managers need to be aware of the costs of BIM implementation as well as the expected savings. BIM value is calculated by subtracting the costs of BIM from the money saved through increased productivity. The value of BIM can be calculated by a variety of key performance indicators, or KPIs. This will allow for intelligent decision-making regarding BIM adoption. The construction industry has claimed various figures about how much BIM could save them in their case studies and testimonials. The numbers vary because of the different KPIs used to determine ROI and productivity improvements (Hong, 2019).

2.2 Levels of BIM

There is currently no clear definition of what BIM is. BIM is defined in many ways by professionals, which makes it difficult to discuss. There is no consensus among the many organisations that have attempted to define BIM. Many aspects of the model are the same, but how BIM impacts the processes varies. In this thesis, I use the NBIMS's (National Building Information Modelling Standard), which is a comprehensive description of BIM. (UNESCO, 2023)

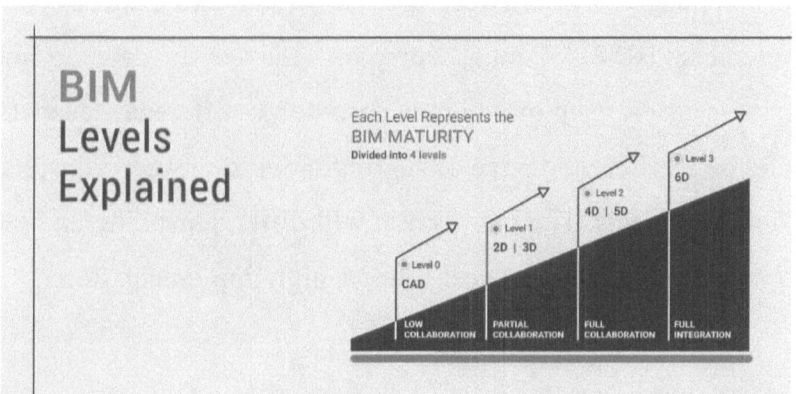

A smart representation of a building. The information is meant to be kept by owners or operators for the entire lifetime of the building. Collaboration process: Covers business standards, automation capabilities, and interoperability to sustain information use.

BIM is often defined as having a single model that stores all the data. Many professionals find this single model cumbersome and will require it to be combined with another type of data storage. The geometrical model could be kept simple and coordinated through a database. It is unlikely that the single building information models will ever be realized. The AEC industry is a diverse industry with many different companies, each with their own unique characteristics, such as size, profession and experience using BIM. A business case should achieve specific goals, taking into account the requirements and company characteristics. There is no standard business process to implement BIM. A single company may also develop multiple business cases, each of which is based on a different scenario. There could be a different degree of sharing models between consultants on different projects. The initial costs will differ depending on how BIM is implemented. Highest cost follows high implementation.

2.3 Benefits of BIM and adoption

BIM has been cited by some as a means to increase the AEC industry's historically low level of productivity. Kiviniemi says that studies show the AEC industry has a much lower productivity than other industries. Some even indicate that it has decreased in the past 40 years. The lack of information and the need for redundant systems are to blame.

BIM is not only a way to improve productivity in a project but also an effective tool for improving performance during and after the project. Ding et. al. state that the adoption of this technology is most advantageous in three areas: digitalization, purchasing and benchmarking. They also struggle to capture useful data from the facilities and their associated activities. BIM is a great tool to help support the Facilities Management System by providing a platform for storing information throughout the entire lifecycle of a project. It can also be used as if it was commissioned, or maintained. (Oti, 2021)

Despite the fact that the benefits of BIM are clear, in terms of increased productivity and

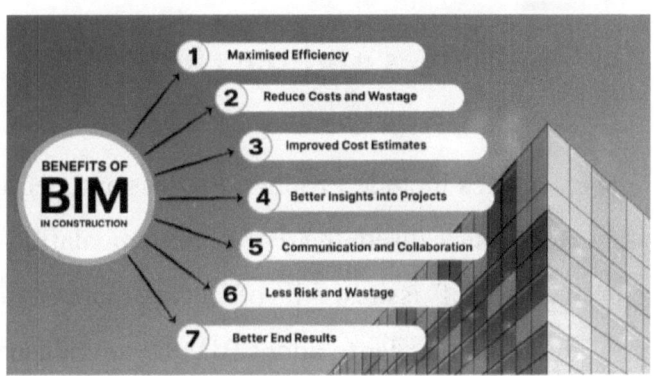

Figure 7: Benefit of BIM and adoption

many others, adoption is slow. AEC is known to be slow at adopting new technologies. BIM can have a significant impact on the way work processes are carried out in projects. BIM can only be successfully adopted if these processes are changed. This fragmentation is problematic because the change can't be implemented by a single actor, but rather must affect everyone involved. BIM adoption will focus on innovation, integration and collaboration in conjunction with significant cultural changes within the industry.

2.4 Challenges and barriers

The categories include Product, Process and people. It is not just one issue that has caused the BIM implementation to be slow, but a combination of issues. All of these factors must be taken into consideration in order to adopt BIM on an industry-wide level.

Ku and Taiebat claim that different BIM programs are incompatible, so data from one must be transferred to another, rather than being shared. This is contrary to the purpose of BIM. It has been a hindrance to the use of BIM for some stakeholders, as they believe that the re-entry information will negate the benefits BIM can bring. There are also few BIM support software solutions for small and medium-sized enterprises. Due to their unusual data, legal issues have arisen regarding who is the owner of various BIM model information, including designs, manufacturing and analysis. (Liu, 2020)

Chan and other studies have shown that the lack of skilled workers has been a significant roadblock in BIM adoption. Aranda Mena and colleagues claim that when there aren't any workers who can advocate BIM adoption, it is easy to discuss its adoption because there aren't individuals capable of executing it. Sebastian also argues that BIM is not designed to be able to integrate such cutting edge technology, making it difficult to use it in projects of this nature due to the lack of coordination and poor preparation of contract procedure. It is unacceptable if the project's coordination is poor and that the process is poorly stated. BIM must be incorporated into the contract at the beginning. BIM is not widely used because of the tweaks that are required before it can be implemented. (Marsh, 2019)

BIM requires fundamental change in enterprises. A coordinated set of modeling techniques and a common model for the construction phase, as well as a shared building model at the planning stage are essential. Some specialists do not see BIM as an alternative to traditional building methods, perhaps because they don't think there are any problems. Nigeria, like other countries developing in Sub-Saharan Africa hasn't passed any legislation to encourage BIM adoption or education. The UK, China and USA are more advanced nations. The government, as the primary owner of a project, is expected to lead the way in BIM implementation. However, this lack of a regulatory structure, especially due to the absence of any economic benefit, leads to a waste of labor and transportation. Other private sector organizations have been discouraged from taking BIM initiatives   seriously.

Construction firms operating in emerging markets face additional challenges, including a lack of involvement from customers and stakeholders, inadequate BIM expertise, and the absence a BIM advocate. The BIM adoption/use conundrum is centered around questions of ownership and patent rights, responsibility for the design and who should administer and build BIM. Gamil and Rahman cite financial constraints, lack BIM knowledge, poor BIM methodologies, lack BIM benefits and awareness, as well as a lack government backing, among other challenges. BIM adoption can be hampered immediately by "geographical location, the economic state of the country, government policy and desire to make a change".

The BIM tool development and tools themselves are at the root of many issues that limit the use of BIM within the AEC industry. The main issue is the lack of interoperability of different BIM tools and programs. However, this isn't the only area of concern. BIM is not always able to meet the demands of users.

This change will require that the way we work is fundamentally changed. In order to develop the basic BIM model of an integrated single model that covers all phases of project development, multiple participants must be able to collaborate. Multidisciplinary projects supported by BIM must have a multi-disciplinary approach. This requires a change of roles for clients, contractors and architects, as well as new contracting relationships.

BIM's new approach to the lifecycle of building projects requires a collaborative approach that is more integrated. To ensure sustainability, the end users, facility managers and contractors, as well as specialist contractors, are required in the design and planning phases. The changing role of project participants, and the new procurement techniques that have followed BIM implementation are reflected in this.

To ensure the success of BIM, all stakeholders must work together to update, insert or modify the information within the BIM model at the various stages during the lifecycle of the facility. The model is designed to reflect and support the different roles played by all stakeholders. The model should be a representation of a facility which can then be passed from the designer to the contractor, subcontractors, and finally to the client. The BIM model evolves as the project moves forward. It will serve as a model central for all project stakeholders. The client can evaluate and compare the suggested design changes or solutions to their requirements throughout the entire project. (LB, 2022)

The BIM model brings with it a significant change to the way work is done. This can lead to many problems. There will be many questions when all stakeholders insert, extract, update or modify the data in the BIM models. The risks that can arise from the use of a model in a way it was not meant to be used are another problem. The results can be disastrous if, for instance, information about the base structure of the concrete is included and then used to procure a curtain wall that it may not have been designed for. In order to address this problem, models now have labels such as "for reference" or other disclaimers of accuracy. This is because designers do not want to risk their usage. This type of disclaimer will make it less likely that other participants in the project are going to adopt this technology, and the designers may have a harder time receiving compensation from the potential efficiency savings associated BIM.

It is important to understand that the speed at which an electronic design may be altered and modified can have both positive and negative effects. When is the final design for the contractors to bid on? Or do fast design changes with immediate notice require the same fast agreement and knowledge of the changes as well? To avoid disagreements it is important to create a project definition and protocol on how information about the building should be exchanged. Otherwise, all participants may not be at the same level in discussing what was offered or what had been accepted.

Gu and London claim that one of most important process issues is that projects won't succeed if business model changes aren't made in a manner that suits the diverse industry needs. According to how BIM is maintained and managed, additional agreements may be required in order to guarantee data security and the confidence of users.

New technology offers designers and builders new possibilities in terms of process, which go beyond the well-known business protocol and norms. Bernstein and Pittman discuss the issues that need to be addressed when transforming the existing transactional business process into new standards. BIM will allow for better information flow and more seamless connections between processes, but will not resolve the challenges of business. Different roles within the supply chain for building are associated with different obligations, rewards and risks. These factors will also change with the introduction of BIM. This new technology must first be adopted by the entire enterprise.

Tse et. al. suggest that the main reason why architects do not adopt BIM is because of the lack demand by clients and project members. In their survey, a large majority of respondents agreed that existing entity-based systems can meet their design and drafting needs. Their study has also shown that BIM's capabilities to handle documents are underdeveloped and there is a lack of awareness. (Tse, 2019)

AEC is not interested in changing technology because there aren't enough case studies that show the financial benefits of BIM. Those clients that are using BIM successfully may be reluctant to share their experience. Some of the most prominent property owners who are moving towards BIM have a public organisation that has a responsibility to share their knowledge in order to help support this change. The more BIM cases that are published, the easier it will be to convince property owners and contractors of BIM's benefits.

To ensure that BIM is implemented successfully, all actors in the industry must be aware of its potential benefits. To take advantage of these BIM benefits, it is important that all those involved in BIM are skilled at its usage. The third set of obstacles to BIM adoption are related to the people who actually work with this new technology, and the need for new roles and new training in order to facilitate the transition. The adoption of BIM is likely to affect both the work processes and the relationships between the actors involved. In the form of model managers, Rizal and co. present a new role for construction   projects.

A BIM model supervisor is needed to coordinate the new technology. Model manager's role will include dealing with both the project and the system. She will maintain and provide the technological solutions needed for BIM functionality, she will manage the information flow, and enhance the ICT abilities of other project participants. The expert will need to be knowledgeable about both ICT as well as the construction process. The role of this actor is not to be involved in decisions regarding engineering or design solutions or organisational processes. Instead, it will focus on ensuring that BIM is used successfully by all parties.

It is important that individuals who adopt BIM are adequately trained to use the technology. This will allow them to adapt to a changing workplace. Aranda Mena and others have also emphasized the importance of good training in their interviews. To ensure the success of BIM, all members of the team must have the necessary skills to use BIM within their field. Yan and Damian conducted a similar study and found that many companies who didn't use BIM thought that training was too expensive in terms of time and resources. They also argue that training costs are the biggest barrier to BIM adoption. The main reason for making decisions is to make a profit. AEC industry is not interested in BIM because there aren't enough case studies that show the financial benefits. Many architects have a social or habitual resistance towards change, as they are comfortable with the current tools and processes of their design and construction. They are also sceptical about this new technology. Some actors do not want to learn how to use BIM tools.

2.5 Challenges on legal level

Ownership of the BIM model is a major issue with BIM, a brand new construction document. Ownership of the model may be a concern for the project owner, who pays the designer. However, other members of the project team might have contributed property data that needs to protected. Each project has its own unique circumstances, so the question of who is the owner of the model needs to be answered differently. Discussions about licensing may arise in connection with these questions when members of the project team other than the designer or owner contribute data to the model. The issue of contractual responsibility for errors in the BIM models will need to be discussed. Updates to BIM models and ensuring accuracy are a risky process. Thompson and Miner state that indemnities and limited warranties for BIM users, as well as disclaimers and limitations of liability on the part of design teams are important negotiation points before BIM can be implemented.

Estimation will be more accurate if you can extract cost and time calculations from the model. This raises questions about the responsibility of interoperability among different programs. Integration between actors is fluid when data is provided in the same program. If data is missing or data has been delivered by different costing or scheduling programs that are not interoperable, then a member of the project team must enter and update the data. Thompson and Miner believe that the contract must address the issue of accuracy and coordination in cost and schedule data.

2.6 Interoperability

Due to the nature of construction, it is important that all parties involved in the project work together and exchange information. This exchange used to be done through drawings and documents. However, with the adoption of BIM there are new requirements to make sure that information is exchanged effectively. BIM is more than just a design tool, it's an interface that allows information to be exchanged between the different phases and actors in a project. The different players are using tools that come from various vendors, or those specialized to their industry. The lack of interoperability or inadequate BIM tools can make it difficult for actors to exchange information. BIM development has progressed to meet the needs of different professionals. The result of this process was a variety of programs which did not work well together or with project management software. Interoperability within existing BIM software and the creation of accurate multi-purpose models are two major challenges facing technology developers.

BIM models are accepted to be built using the Industry Foundation Classes (IFC), defined by buildingSMART. IFC represents an ambitious effort to achieve model-based integration. The information is not only limited to the geometrical characteristics of objects but includes metadata relating to the other features of the building.

- File-level interoperability is the capability of different software to exchange files.
- Interoperability at the syntax level - this covers how different tools can successfully parse files, without error. It also includes the capability of different tools to work together without error.
- Interoperability at the level of visualization - this is the capability for tools to display the model correctly.
- Interoperability at the level of the theme - The ability of different tools to understand the meaning and significance of an exchanged model.

Steel et al concluded that IFC had achieved relative interoperability at the level of files and visualisations within a small subset domains. It is particularly notable in the architectural design domain. It still has challenges in situations requiring sematic interoperability. This is true even when the use of sematic interoperability is expanded to encompass more subdomains. Interoperability in the construction sector is a problem due to the wide range of projects. These can be anything from simple single family homes to airports. IFC's interoperability has suffered because of this breadth, as no tool can implement all its languages. Interoperability in the AECindustry is important because of its fragmented nature and collaboration. BIM offers many advantages over CAD but sharing intelligent building information, which is a critical aspect of BIM, is crucial. Interoperability is a key challenge to overcome in order to maximize the productivity benefits and quality of design that BIM offers. (Xu, 2019)

This literature review was conducted to gain a better understanding of BIM, and its barriers. The literature review was used two-fold, to confirm my prior knowledge of the topic and to compare the empirical data. My literature review is focused on BIM, BIM benefits and BIM barriers.

There have been several studies done in projects that BIM was implemented. These case-studies provide useful secondary data to support the thesis. These cases, however, are not generally located in Sweden and may not therefore be representative of conditions in this country. These are the cases that were studied:

KTH Campus Utbildningshus- Akademiska Hus plans to build new buildings on the KTH campus in "Valhallavagen". The project is a BIM implementation and can be used as an example. The project is still in its initial stages, so it will be fascinating to learn how BIM issues are handled and the decision-making process.

Nya Karolinska Solna – Construction of the hospital "Nya Karolinska". This project has high expectations for the implementation of BIM, and these will be maintained throughout the lifecycle of the project. This project is a good example to use when studying how BIM could be implemented in the Swedish context. The project at KTH Campus has been much more advanced and other questions will have arisen and be addressed that will interest the research. (Ren, 2021)

Gu and colleagues concluded from their research that views about BIM depended a great deal on each actor. The profession of the individual and the size of their firm can have a significant impact on the perception of BIM and the way they want to use it. Larger firms that are likely to be more involved with large projects prefer flexible tools for 90ustomizing the project environment. Smaller firms on the other side, prefer intuitive environments. Professionals in the AEC industry want BIM to include all the capabilities of CAD so that they can continue to benefit from them. There is also interest in new capabilities BIM can provide. They want BIM tools to have new capabilities without removing older ones.

The expectations of BIM vary depending on who is using it. BIM is viewed by designers as an enhancement of CAD. Project managers, contractors and other professionals see BIM more as a document management system that can extract and analyse data from a CAD program. This survey suggests that, although BIM developers strive to incorporate both aspects, current BIM applications do not meet the requirements of either.

The result is that different stakeholders have different demands and expectations on BIM and do not share a common understanding. Most studies focus on BIM's enhancement of current CAD technologies, but do not highlight its document management capabilities. While BIM is being developed to help with facilities management, it lags behind BIM's development in design phases. The lack of interest from non-design professionals in adopting BIM has been a major factor. Gu et. al. argue that user-centric BIM development must be inclusive, as successful BIM adoptions require collective contributions and participations from all project stakeholders.

3.1 Case Study 1

Around 1100 students are expected to arrive at the KTH Campus in Valhallavagen by 2015. This will create a demand for additional facilities. The parts of KTH that are currently located in Haninge, will be moved to the campus. Architectural school will also get new facilities, though they lack the lecture rooms that are currently available. KTH needs new facilities and is planning a building in a space   that   is   suitable.

The law in Sweden prohibits the universities from owning their buildings. Instead, they are required to rent them. The facilities on the KTH Campus at Valhallavagen, are currently rented by Akademiska Hus. This is a Swedish state-owned company that owns the majority of the university and college buildings in Sweden. KTH can't build the necessary facilities on their own, but instead must negotiate with Akademiska Hus to construct new buildings. (John, 00)

KTH needs new facilities to conduct its education. There are many requests for new constructions at KTH, mainly in the department of Civil and Architectural Engineering. The following are some of the first requests made regarding the building:

Create an environment that encourages innovative thinking and is flexible. Use the facilities and technology to enhance the educational process. KTH should be innovative in its approach to sustainability and the environment. KTH also wants to build a building that demonstrates the university's position as an advanced technical university that provides state-of-the art education. KTH can make demands to Akademiska Hus because they are the tenant of this project. Akademiska Hus, on the other side, can demand that KTH pay higher rents if the demands are linked to higher costs of construction for the project. Discussions are also interested in how these costs will be allocated to all of the facilities that KTH rents through Akademiska Hus. The higher rents after KTH's expensive request are only for the planned new facilities. This will reduce their use and value because they   cost   more.

Akademiska Hus owns this project. Both KTH and Akademiska Hus see this as 'a prestige project' to demonstrate their ability to build buildings of high-quality. Rents will increase because KTH leases its facilities to Akademiska Hus. It creates an interesting scenario when KTH makes additional requests.

KTH is ambitious in its vision of how to use the new facilities as an active component in education. The new building is an excellent opportunity for KTH students to use practical examples in their engineering classes. This project is intended to be used as an example for how construction projects are carried out. KTH is also interested in equipping the building with sensors that will provide data about how it is actually performing. Parts of the requests have received funding. Filming of the project will provide information on how meetings and procedures are carried out in real projects. The material can be used for research and possibly educational purposes. The building will also be constructed with BIM. KTH wants to use these models   for   education.

BIM has limited examples in the real world, and it would be helpful to have more practical BIM experience. The material can also be used for further research in order to develop technology. KTH has requested that the BIM model should be made available to KTH in accordance with the agreement. KTH must have access to all the information in the BIM model for it to serve its purpose. On the other hand, KTH has no desire to modify the model that is linked to a building as they will not be making any modifications to it themselves. KTH wants access to the data to analyze and educate, but does not want to add more data or change the data. This request for BIM is not directly related to performance. The demand for BIM does not lead to any demands to lower rents, or other compensation. It is only a request to have access to the models.

Akademiska Hus uses BIM for most of its new-construction projects. Most projects involving refurbishment of existing buildings or the construction of additions to them are not BIM-based. BIM adoption is influenced by both the specific project characteristics and the project manager. BIM-experienced and interested project managers at Akademiska Hus are more inclined to include BIM into their projects. Also, the type of project has an impact on this decision. BIM is not used for small projects involving renovation or reconstruction. Existing documents and drawings as-built are not usually compatible with BIM in such projects and the creation of BIM model for existing structures is quite costly.

Akademiska Hus has ambitious goals for BIM in projects that involve the construction of new buildings. Akademiska Hus developed a BIM guide to instruct their project managers how BIM can be used. This manual states that the purpose of BIM in a project is to improve the communication between the parties involved and to exchange information. This will reduce costs and increase efficiency by reducing the number of errors. Akademiska Hus sees BIM as the new CAD that will be integrated to make the processes more efficient. Akademiska Hus focuses on BIM's informational part and distinguishes between BIM, 3d models and BIM. BIM goes beyond 3D design; it is about the connections between objects, information and models. BIM models can be developed more quickly if project owners, contractors, users and designers are involved early.

Akademiska Hus wants to also use the models for facilities management after the project has been completed. The BIM manual explains that models are adaptable to be used by facility managers in the future. Currently, there are problems in using BIM as a support system for facilities management. The models do not have the tools necessary to make the most of their information, and as a result they are not integrated into the ongoing facilities management activities.

Akademiska Hus hasn't deemed this a major issue at the current stage of its general BIM adoption. Project managers with BIM experience are those who have the most knowledge and interest in how to use it. Consultants and contractors are generally familiar with BIM. Changes in BIM tools and practices are mainly related to how models can be collected, stored and used later.

2.3 Case-study 2

This is about Nya Karolinska solna.

Nya Karolinska Solna is a complex and large construction project that began in the summer of 2010 and continues until fall 2017. The hospital will have a total area of approximately 320 000 sqm. This hospital will be built and managed by a PPP that will last until 2040. The PPP involves Stockholm County Council, Swedish Hospital Partners AB and Stockholm County Council. This project company will be responsible for the construction management and facility management of NKS up to 2040. Interviews with actors have led to the results of this case study. (DP, 2018)

Stockholm County Council, the local authority responsible for the healthcare of the Stockholm area. It is therefore the one who orders the construction of new hospital buildings. LSF, the facilities management of the County Council Board is responsible for building the hospital. The subunit NKS Construction is responsible as the client of the entire project. This includes both construction and a 30-year maintenance contract.

Karolinska Universitätshospital is in charge of developing demands for medical equipment and information technology. The Council Board Facilities Management is in charge of purchasing the necessary technical equipment as well as the interiors. Swedish Hospital Partners AB, a project company owned by Skanska Infrastructure Development and Innisfree in equal shares, finances and executes the project. Skanska will be the one to perform the actual construction through the design/build method. Coor will be responsible for managing the NKS facilities from completion to 2040. Project company is funded by Swedish and foreign commercial banks. They are also responsible for ensuring that construction proceeds according to the plan, and delivering their payment before the County Council makes their payments.

Public-private partnership (PPP), in this case, is intended to provide a high level of predictability and dependability for society and taxpayers. Stockholm County won't pay annual payments to the company until the project has been completed. This solution creates strong economic incentives for them to finish the project in time. PPP includes both construction of hospital facilities and facility management. In turn, the project company must build the hospital facilities according to contract.

It is the project company's responsibility to maintain the performance of functions as set forth by the client. The yearly payment is reduced if the project company does not meet this obligation. The project company is then motivated to build facilities that will last over the long term. PPP allows the County Council, because of its fixed price and agreement to manage the hospital each year. The County Council transfers the financial risk to the project firm. In a large and complex project such as NKS, the County Council has determined that PPP offers significant benefits. A planned restructuring of the highly-specialized healthcare system in Stockholm County makes the completion date of particular importance.

The construction phase will last from 2010 to 2017, when all the new buildings are expected to be finished. The County Council pays a total amount of 9.868 million SEK in this phase to the Project Company, which is divided into three categories.

- The payments based on the results will continue to be made continuously. They will total 7 321 millions SEK and correspond roughly half of construction costs. Banks and the owners of the company that is undertaking the project will provide the rest. County Council guarantees that it will not make payments until the banks confirm that project process is going as planned, and they have made their scheduled payments.
- After the construction phase has been completed and inspected, a one-time payment of 1 386 millions SEK will be made.
- Payments per year - total 1 179 millions SEK. This payment will cover the costs of operation and maintenance for parts of the hospital which are put into operation prior to the completion date.

The project will end when the NKS Hospital is operational and the annual payment for the project company is calculated in the 2010 monetary value. The County Council will provide the project company a total amount of 52 billion SEK, which is equivalent to 27 billion SEK in 2010 value. The payments are for the construction, financing, operation and maintenance of the services related to the building from 2010 through 2040.

The owner of the project (Stockholm County Council), introduced BIM to the project in the contract signed between the two parties. This contract contains definitions for what is to follow with BIM and the models generated. This project has many models that follow the life cycle of the project. The life cycle will last until 2040, when the contract ends. At that point the model will then be given to the County Council. The contract between Stockholm County Council, the project company and other parties does not include any requirements for open formats. Interoperability issues will be transferred to the company that is undertaking the project. The model is object-based and attributes are either linked or directly included in it. It is possible to define spaces and connect them to equipment that belongs to those areas.

There is no requirement in the contract for BIM open formats such as IFC, to guarantee interoperability. Skanska is responsible for combining the data from designers. Skanska does not have to convert any files because the models are delivered by the designers in Navisworks. Skanska, as main contractor, is responsible for documenting the as-built model. The models will be provided to the facility manager for their ongoing operation and management. De design consultants can also use these models to support their work. The large project size made it impossible to gather all the information into a single model. It is impossible to use a model of this size due to the lack of computing power. Therefore, it was broken up into several models that contain the same information as a master model. BIM managers have also been trained to be present in the various actors, ensuring that the models are used properly.

BIM was brought to the attention of this project by the Stockholm County Council, and incorporated into the contract signed between the council and the company. One person was particularly knowledgeable about the benefits of BIM. She worked to include BIM into the contract. However, the goal of adopting BIM is not clear in this case and there is a lack of a shared understanding about how the models are to be used. Coor's ambitious BIM use is a result of combining the need for BIM with informed and enthusiastic individuals within that organization.

The tenant, Stockholm County Council (Stockholm County Council), has requested the use of BIM for the project. This will be followed by the maintenance and operation of NKS. BIM, on the other hand, theoretically allows for greater reliability when it comes to the maintenance and operation of facilities. This will be beneficial to the end user in this instance the tenant. BIM implementation can also reduce total risk in the maintenance and operation stage of the project. This could lower the costs to the tenant.

This project has been delivered using the Design/Build delivery method. Skanska is the contractor who takes on both the design and the construction. Skanska was responsible for collecting the necessary information to create the BIM model as part of the contract to produce BIM. Due to the size of the project and the complexity of the buildings, it is a large-scale undertaking. To complete the project, many designers contributed information. To allow these designers to share information, they have made agreements about the programs that are allowed to be used.

The agreements guarantee a high level of interoperability in the project. In contrast, there have been no demands for the project to use open standards such as IFC. Skanska received the completed models in predefined formats to guarantee interoperability. This stage involves the creation of different models for the design disciplines, which Skanska then combines to create the BIM, with the data defined in the contract. The adoption of BIM did not change the process of design in any significant way. Instead, it was used as a tool for communication and coordination between different disciplines.

Skanska utilizes BIM models to a lesser extent, but still not significantly. Mostly, the models are used to locate construction sites. The model will be updated continuously during this phase to reflect the correct building and all its components. The model is then handed to the facility manager.

BIM is primarily used in facilities management in NKS projects. Coor Service Management has very high goals for the use of models in its work process. The contract between Stockholm County Council (SCC) and the project firm outlines the use of BIM for facilities management. This is similar to earlier phases in which BIM was used. Coor, as facilities manager has developed its use of BIM. It aims to achieve a much higher level of usage than the minimum stated in the contract. The NKS facility's extensive use of BIM for facilities management has higher goals than any similar project anywhere in the world. The use of BIM models is a high-profile project, and a way to show the value of BIM at these phases of the building life cycle.

BIM, for Coor in his role as facilities manager is mostly a method of managing data. Information is not as useful when it's stored in traditional folders. BIM, in contrast to these systems, allows databases to be linked directly with the model to make it easier to access the information. The facilities manager is very interested in systems that support data management and ensure a good management of the information. BIM is a way to connect the huge databases that are created by the various actors in the construction and design of a building with the facility management system. The NKS project gains two specific benefits from this:

A BIM model as built will provide the facility manager with the necessary information to understand what is in the building, and how the maintenance should be done. This knowledge can be very valuable to both the operating staff and the maintenance team, as it gives them the correct information about the types of materials used and the functions of the facility. It also tells them how the supplier recommends that the building should be maintained. The facility can be more efficiently used if you know how to use the various spaces and for what purpose they were designed. The facilities manager will find it very valuable to be able document the performance of the facility. Databases linked to BIM models will allow for the storage of data related to facility performance and evaluation of changes to be made in order to enhance its function.

Coor says that BIM allows for the collection of hard data about how facilities perform. This was previously done through a softer assessment by professionals who worked with facilities management. This ability will directly affect the cost of maintenance. For example, if the expected lifetime of the elevators is 15 years, but they aren't working properly, then data about the cost that such issues might incur can help determine if the installation needs to be replaced before its life expectancy ends. The costs of maintenance aren't the only ones that come with faulty installations. Coor may be penalized if the elevators do not function within certain parameters. Coor will be able to make more accurate estimates about the performance of its facility and the improvements needed by collecting and storing the information. The maintenance costs will be reduced and the risk of penalties due to insufficiently working installations is also decreased.

BIM can be used to deliver traceability into an information system. The databases can be linked with the model to determine both the specific quantities and the locations of objects. Maintenance personnel can then identify the exact location of faulty installations, and compare it to product data stored in the databases that are linked to the model. It is also possible to provide services related to specific activities in the facility. It is important to track the movement of certain items in a hospital to prevent infection. This project aims to deliver such information.

BIM can provide tenants with new services in addition to developing and enhancing the functionality of the NKS facilities management system. The BIM model with its linked databases allows Coor to offer many new services. There are discussions on how the processes could be improved, but it's technically possible. The model can help with booking rooms, for example. This allows administrative staff to search and find the best rooms to accommodate patients without having to call different departments. The cleaning of the rooms can be linked to their use rather than a standard schedule.

Rooms could, for example, be immediately cleaned when the hospital staff registers them as being empty and registered by cleaning staff as ready to use. It is possible to link the amount of usage with cleaning. For example, automatically register the toilets as needing cleaning after a certain number of uses. Although these possibilities may sound far-fetched, they are a reality and possible with today's technology. These new opportunities will require some adjustments in the way we do things and new systems for facilities management. However, once implemented these improvements in services as well as the overall quality of the building can be seen by the users. (Xu, 2023)

Different design consultants find that they are most comfortable and productive using tools of a different kind. To ensure these actors could remain productive and still interoperate with each other, it was important that the design consultants agreed on what tools to use. In order to be able use a tool, there had to be sufficient interoperability between the tools that were used by others. The solution could have restricted some actors by providing them with tools that were not interoperable with other actors, but it adequately addressed the critical issue of interoperability. Skanska insisted that the design elements be provided to them as a certain format. The design consultant was then responsible for any conversion issues. This allowed Skanska to have perfect interoperability. The model was developed to satisfy the requirements of the contract and requests made by Coor, the facility manager.

Skanska's representative said that the BIM implementation has not resulted in a significant change to work processes. Designers are using many tools that would be close to BIM, but they haven't been combined and exchanged into one collective model. When comparing NKS with other projects, the difference is in the consistency of how design consultants have uploaded their models to the master models that are managed by Skanska. Skanska's contractual obligation to create BIM models has created a strong incentive for them to complete these models correctly. BIM will have the greatest impact on the NKS life cycle in terms of changes to processes. This is mainly due to Coor's extensive use of BIM during the phase of facilities management. In the literature, there is not much evidence of the collaborative working process.

Chapter 3: BIM and construction project management

A project can be defined as a temporary endeavor to produce a unique result, product or service. A project is not always temporary, especially in the construction sector where they can last for years. However, a project does have a beginning and an end. Even though a building's structure is often similar, they are all constructed in different conditions and therefore are unique.

Project management can be defined as the process of applying relevant skills, knowledge and tools to an undertaking to achieve its objectives. Projects have five main processes: initiating, planning, executing, monitoring and controlling, closing. Projects have different phases and processes which need to be managed with caution. These processes must be managed correctly: Identification of requirements; Addressing stakeholder needs, concerns, and expectations during the entire project process; Balancing project factors such as Scope, quality, schedule, budget, resources, and risks.

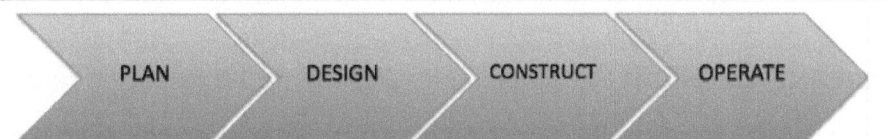

Figure 8:construction phases

If one of the factors is changed, it will most likely affect at least another factor. If a stakeholder lowers the budget of a project, it will have an impact on all other factors such as quality, scope and budget. You also have to consider that stakeholders may place different priorities on different factors. The project management plan that was set up at the start of the project must be constantly updated throughout the lifecycle and changes need to be managed correctly.

Building Information Modelling, for me, is basically a way to represent all of the components within a building on a three-dimensional basis. All the attributes of those components are identified. The building information model can be easily updated and will leave a lasting legacy for the entire lifecycle of the structure." Andrew Hallworth (AH). You create 3D digital models for a building project and then add information to the objects. It is used to both coordinate and communicate between disciplines within a project. It is an ongoing process that takes place throughout the life of a building, starting with the initial design phase and continuing through the construction, management, and demolition phases. When you create 3D geometry in design software and then transfer the information about quantities, size, and number to another database, where tech-information is added, this is BIM.

It is then possible to share this information. The data becomes accessible in another form and these objects are only represented in the model in case holders. When I see this, it's when you started to use BIM. BIM is an excellent investment. The contract stated that the BIM project was one of the projects the county council had acquired. Ulf is involved in the first BIM project where the entire team draws 3D. Early in the project planning, it was decided that the BIM model would be handed over with documentation on facility operations and management. Peter is astonished by the fact that they were able to be so foresighted in this project and set such a high bar. The rotation time in such a large project is long, so it can take up to 10 years for the requirements to be met. In BIM, a lot happens in 10 years. Peter thinks that BIM integration is at such a high level that, when the building finishes in 2017, it will still be on the cutting edge of the industry. In a smaller project, it is possible to experiment and push the limits of what's technically feasible. It is also important to consider the processes that are used when doing this on a larger scale, such as NKS.

3.1 BIM and environment

This project has made a significant investment in the environmental profile. The material list is linked to the BIM model. This has been chosen as a way to incorporate BIM into the environmental requirements. This allows you to highlight an item in the BIM model and obtain information on the materials that make it up. The material list can be highlighted to find out where a particular material is located in the BIM model. This project is unique because of the extensive international collaboration that exists between Sweden and United Kingdom. Skanska, a U.S.-based company, has helped the NKS in providing an effective platform for BIM within the Facility Management.

3.2 BIM, compromises and challenge

In the project agreement, it is unclear what the use of the model means from an operation and maintenance point-of-view. This has led to a lot of discussion about how people interpret the definitions. The result has been confusion and challenges throughout the entire project. From the Swedish Hospital Partners, the contractor hired to build the building to Coor Service Management to manage the FM. For example, the agreement does not specify how information will be connected to the model. There is no list of requirements for the level of detail to be included in the model. This is a very dynamic process. Together with FM, all disciplines must find the appropriate level   of   information.

The effects of BIM on this project are difficult to understand. Peter doesn't believe it will be possible in the NKS Project to generate a return on investment for BIM. There is no clear line between design and BIM. What are BIM costs, and what traditional planning costs are? It is therefore difficult to determine the exact cost of BIM. It is difficult to determine the exact amount of money invested in BIM. BIM can be used to get a model estimate in an hour rather than a day. It is a positive image on a local scale, but it's important to compare these advantages to the amount of money invested during the design phase in order to make BIM possible. It is not possible to compare the amounts invested in BIM because there are no exact figures. Peter believes that the BIM investment has been worth it.

3.3 Expert people about BIM

Chris believes that it's too soon to tell if BIM will be a success for the participants of the project. Human nature is to resist new things and not see their benefits right away. No one wants to be a Guinea Pig. After the initial involvement in BIM, the people start to see the benefits. Only a few people had experience in BIM before the NYA Royal Adelaide Hospital (NRAH), project. BIM is more than 3D modeling. Health planners, who were used to BIM requirements of 30-40%, had the most experience. They were used to using planning software that had codes. Then they extracted these codes. This is similar to a data management system. The data was extracted by calculating the space data sheet based on the machine weight. Health planners were therefore familiar with the BIM processes. The architects, on the other had little experience in BIM. There were people who were familiar with BIM because of the diversity of nationalities in the project.

There is an enormous need for training in NRAH. Currently, 18 contractors are being trained weekly. However, this number will soon reach 100. Not everyone is trained in BIM. It is likely that the worker who pours concrete at the site does not need to know about BIM just yet. People don't always adopt the new technology as anticipated. The project team had thought that some contractors would adopt BIM naturally because they were familiar with technology. However, the opposite has been true. Instead of integrating BIM, they have focused on reasons to not use it. People who thought they would never use BIM are now very adept at it. The fact that the BIM implementation was one of the very first large projects to be implemented in this state and it took four years to reach financial closure is important. Chris shows people how BIM has been used for over 20 years by showing them previous projects that have proven successful. It is natural to doubt the success of something new. The problem is, after working together for a project and understanding BIM, the majority of people don't work together again on another project. The second project is necessary to see any real benefits.

This project has taught us a great deal. The contracts were not perfect. Chris says that the project follows a similar pattern to many major projects in Australia. Industry needs to collaborate more. This problem will continue until there are consistent documents for this type of construction project. Chris has been doing his best to fix the problems, even though the contract was signed many years ago. It is an uphill struggle to get contractors to do anything after signing up.

Chris says that 5D is really a business and it depends on the company to organize their entire business workflow. The NRAH is now using their own database to do 5D design, and link costs without relying solely on intelligence from models. Then they add it to their data, and use the outputs for a timetable. They will then include the costs in the schedule. It would have been impossible to estimate the costs of the project if they had used any structured outputs early in the process. They have created all of the health and medical equipment from the very   beginning.

There is limited international communication among suppliers. Some of the suppliers had not heard of BIM, or any of the other software before the team requested that they contact counterparts in America and the UK. Chris asked the question half a year ago and some of his suppliers found libraries in other countries. The architects were already away, creating a 40-lot set of contents. They are working hard in this project to promote BIM, the libraries and national object libraries. But everyone still does their own thing. It is important to have industry-wide standards and protocols. The NRAH Project has created a shared LAN to allow better collaboration between Naviworks files when clash detection is being performed. The NRAH project will use ZUUSE for their field BIM, but they have also tried BIM360 glue. It has always been an issue with the size of BIM files. The software will run slowly if additional parameters to object families are added. The model must be divided into design areas in order to make it manageable.

This project has seen a steep learning curve. Even if they were accustomed to modeling in 3D before the NRAH, with the various software programs available. Working with BIM requires designers to be more proactive and this can result in more work. The designers have to model the part in 3D and include all of the details, such as the dimensions, weights, finishes, etc. This is not necessary if you are drawing the 3D version. The BIM model that they produce will prove to be very useful for the life of the building. As a project it is clear that more time was spent on the front-end but more time saved at the end. Andrew says that the construction industry is resistant to changes, especially trade contractors. They are afraid of taking on risks that they may not be able manage. It is very hard to convince trade contractors that they should use a new product when it comes out, unless the client makes them do it. Only when the momentum is going can you expect to see widespread adoption.

Chris joined the project after the consultants created an execution plan. Everyone was working independently. Chris says it took some time for them to adapt to Chris' "we-approach". It's always about people, because we are all experts and want to be different. Rather than just following the rules, people sometimes spend time arguing why they shouldn't. BIM has a reality that is not liked by the industry and it cannot be managed. BIM makes the process transparent, and relies on discipline. The project team consisted of 60 to 70 Revit users and team leaders. MEP consultants would always struggle, so HYLC brought in the MEP contractors early. Chris and his colleagues are making consultants provide information at a LOD200 level, which provides a much better level than contractors had previously. It also shows where there is a lack of engineering. Client User Groups were held for 14 months to develop the initial brief.

Although the project is now closed, it still involved over 450 sessions because the brief wasn't clearly communicated. Chris says that each dollar spent on hardware is also spent on awareness and training, due to the high turnover of staff in large projects. When bringing new employees into the project, it is important to introduce them in a structured manner to get them familiar with what's going on. This is a problem that is constantly being faced by the industry because it isn't used to having discipline. Revit's data is huge, and it can be difficult to keep up with. It is important that people understand and manage technology adoption.

It is essential that the consultants agree on the exact model and how the hospital will be broken up. It sounds simple, but the engineering team prefers to build movement joints. Service engineers prefer to work floor-by-floor. The architect wants to go by department. It's all about how they do it. The first point is that you need to agree on a way of working when using BIM for a big building. This will allow people to work in the same manner and launch models correctly. You don't need to get an unlimited memory and the support for it. In this project, dividing the building was a major issue.

A compromise had to be made. There are different levels of design in NRAH because the consultants, for example, take the coordination of the services to a high level. Andrew then goes to trade contractors and shows them what they are going to actually manufacture and install. Trade contractors then take the model, and manufacturers draw their shop drawings on the model. There are several levels. For the models to be actually delivered to trade contractors, or finish services in this project they must meet a certain design status. The model rotates and comes back to a new level. This is the one that gets built. The model is called the level of design 200, 300 or 400. Everything that goes into it has a specific status.

Chris is saying that everyone is measuring BIM's effects in this project, because the schedule is very strict. It has more to do with design and construction than BIM. This is a PPP for a hospital worth 1,8 billion dollars, with a 2-year design period and 3 years of building. The consultants (results), are monitored by some strict KPIs. Chris had tried to put some KPIs in early on with the BIM, but they all picked it up. One can see that the level of knowledge and output is ten-fold better than when participants started. Chris and his team have been trying to find other metrics, but they've not yet succeeded.

3.4 Pros and cons

BIM helps people gain a better understanding, especially when used to manage data. This type of project is complex and involves many people in various teams working on different tasks but all within the same space. What will the space look like from the architect's perspective? What is the function of the space? Both will have an impact on each other. It is very beneficial to create a central hub of information. In NKS, this would be the model for an architect that connects to various databases. The process becomes easier. You can do clashcontrols easier, for instance, since all of the data is in one location.

Gunilla says that the IFC was used in NKS. She does not like the IFC format. She would prefer a direct solution. They encountered a variety of issues and problems when they used IFC to create the models for energy estimation. The first problem they encountered was information loss. Another was trying to import the 43 structure models from Tekla to Revit using IFC, but it didn't work. When importing the Revit model, every little detail of every screw or welding in Tekla is seen.

Peter believes that the project would have cost more without BIM. BIM has a great impact on the Design process. They can view and co-understand an early BIM version in Design. The clash-controls are a great way to show the benefits of BIM to the general project manager, even if they find it hard to understand how it will help the project. They may think that it's unnecessary or costs too much. Clash-controls can be implemented in virtually every project.

NKS has a simple way for builders to access the information they need to know about the location of the materials. It is not unusual that the BIM model is linked to the digital lists of materials. It would have been possible to provide the list in paper form, but that isn't very user-friendly and difficult to update. They have thus integrated BIM into the environmental requirements. BIM allows the people working on the project to perform two types of analyses: personal and data analysis. Personal analysis is when a person studies the model, understands the building and comes to quick conclusions. NKS is very familiar with personal analyses, even though data sets are not as common. The data sets are analyzed for energy in all parts of the buildings, but cost estimations based solely on models can be done to a certain extent at the individual level. Ulf thinks that today's installations in healthcare are so complex that people planning great hospitals from the 1970s, who could visualize the drawings even without 3D modeling, would struggle to complete this project without BIM. Also, designers today aren't trained in the 3D visualisation that can be projected on 2D diagrams. BIM, therefore, is a must-have tool to complete the design of the hospital's installation, the hardest design to do within that industry.

BIM makes it easy for people to understand the process and what they are doing. BIM offers great possibilities when it comes down to the user interface. It does not generate any new information but it opens up new opportunities for analysis. The main benefit is the way it organises the data so everyone can easily understand it. Some people might have struggled to understand the numbers and acronyms if they had used Excel lists for NKS. If you can view the exact window in a 3D-model, then click the particular window to get the full documentation for that window. In all organizations, communication is crucial. But it's even more important for a complex organisation such as NKS. BIM can be used to enhance communication.

In addition to the paper drawings, lists and descriptions included in construction documents, the database also contains the same data. Even if the information is synchronized with the model, it does not require that one go to the actual model. The database is very useful for quantity calculations, as an example. Peter, a planner and designer manager in the construction industry, feels that BIM helps him follow-up on the design. Peter receives regular updates and it's easy to determine why something wasn't drawn on the model. Every week, he can see the progress of his design and if everything looks good. The transparency is enhanced. Peter can do his analysis when he sees the 3D model. He can get an idea of what the current design is and the future plans. It allows him to reach certain conclusions, and to ask the correct questions of the consultants.

BIM increases transparency, which can lead to an increase in expectations. The client will not be able to identify all clashes if they only see 2D illustrations. People expect that the designs will improve when BIM is introduced, but instead all of these clashes are made visible. It can be difficult to maintain a balance between expectations and actual results. The result may not always meet expectations, but it could feel like one if expectations are higher.

The level of detail in the model is not accessible to everyone. It may not be beneficial for all project participants to have the same level of information when a large-scale NKS project is involved. If the project manager is responsible for the entire project, he may not be focused on the finer details of the project. BIM is transparent in the way it makes design and engineering.

Peter makes a comparison to the arrival of mobile phones. You add your current technology to what is enabled by the new one and then take the total. Today + New Technique = Result. It is not what he meant; new technologies will eventually erode old knowledge and techniques. We thought it was great that we could call anybody whenever we wanted, but the fact is this changed how we lived. People stopped being on time, because it was easy to call in and tell them that they were late. Peters's concern with BIM is this. Implementing this technique could erode existing disciplines in planning and designing a building. We might not be able to clarify the limitations of the various disciplines in the same way. We will no longer set up a structure for how we do design, because BIM allows us to work continuously integrated and manage changes quickly. This isn't to say that it has occurred in NKS, but he does raise a concern as with any development.

It is important to understand the current state of information within the model. This can pose a risk and challenge. The design process is one that involves a lot of back-and-forth. It is still important that the BIM statuses are completed correctly, even when BIM is used. For the name of a design, a system document, construction documents and corresponding program must be completed. There is confusion if you don't. For example, a BIM model was created only a couple of weeks into this project and had an apparent floor that wasn't necessarily the correct floor. Nobody can guarantee anything until the model is delivered with the correct status. It is especially a problem with large-scale projects. This is a risk, as people may take the wrong information from the model.

Chris believes that they would have required twice as many workers if BIM had not been used. The program would have been roughly doubled due to the overrun risk and the lack of coordination. If you can show the finance people how much contingency is on each project, and with BIM's help we can reduce it, then everyone will demand BIM. Contractors, for example, have often had to solve all issues late on in a project. This often results in higher costs than if the issue was solved during the design phase. Andrew says that with BIM, you have to work differently and be more collaborative. The key is to have trust in the other consultants when it comes to pushing around models. There are individual consultants' models and then a coordinated one that all consultants must download.

Chris says that BIM does not only manage risk, but also provides certainty. It makes everything transparent, and the quality is actually demonstrated. It doesn't matter if the model is still changing due to client processes. When they begin to digitally lay out the building on the site, hopefully the coordination and accuracy will be much better than in traditional projects. Chris says that one of the benefits of Aconex is the ability to click an area on the model and instantly have the 2D drawing displayed. This allows the site workers to easily access the drawings. It is part of the BIM field strategy developed by NRAH for the transference of data to the site. The workers can also fill in the digital form for the QA signatures using their tablets. The project team can then use the data for progress reports and link it to the Naviswork model. Clicking on the column will reveal what data is being captured.

Some managers are able to see the advantages, while others do not. They still look for more drawings. The people that aren't hands-on still have issues. The biggest problem is they may not have enough architectural or engineering knowledge. Chris says that the structural engineers have adapted to the new technology well, because they've always drawn things that had to be built. The engineers can grasp the concept of prototyping, while the architects were initially concerned that BIM would limit their creativity. The architects quickly realised that they can link their creative skills with the details.

BIM can be used for all kinds of things, including clash detection and 3D modeling. It is also useful for putting models up on screens to help with understanding during meetings. Andrew's colleagues are able to not only see and find a pipe rendered visually, but they also know all about it; its size, function, origin and anything else. It is easy to see what the pipe looks like and to know what needs to be completed to prevent clash detection. BIM demands a different approach, including an extremely structured mindset. Designers need to be aware of the attributes that need to be entered. Then, everyone has to be glued to the model to ensure that they are all following these rules. It's a similar process to 2D drawing. The layers are pre-defined for consultants to use. If the form and attributes of the layers are not rigid, then they will end up with pieces of the model which do not contain the information required. They end up with model parts where information is placed incorrectly, causing problems.

Architects are especially susceptible to problems, as they're used to 3D presentations. This can lead to architects cutting corners on the design that does not look good. This is not possible when working with BIM. FFE is one of the most significant risks on the project because the NRAH was a hospital. They can analyse the information because they have a BIM Data Management System that manages all of their Revit data. The system allows them to compare each area of the functional brief with the actual brief, and they can also check the FFE items against the original brief. The model can be colour-coded to identify the items the architect added that were not included in the bid.

The consensus on who benefits most from BIM adoption was divided. However, one person commented that "Designers and Contractors" benefit the most as improved accuracy leads to more projects and better collaboration between Contractors. One person said that BIM adoption is beneficial to everyone if there is agreement on the objectives and the approach. Another suggested that clients and end users will benefit the most as they are aware of what he is getting and how he is going to use the 3D model. Some respondents, however, disagreed with this statement. They said that "the client will benefit the most" and added further "but they have to be educated clients". Another responded that, "at this moment, 90% of clients cannot see the benefits and are not able to get them, while most of the BIM is forced on the client" because they don't understand "what BIM means". It is often the case that "middle, senior and client management" are misunderstood and this leads to "poor adoption and governance". There was also a consensus among 50% of the respondents, that the younger engineers and designers are more likely to "get" BIM and understand it. This is because they have more experience on the job. However, the higher you get in the hierarchy the less attached people become. This is believed to be due to the "management's belief that BIM is all about software, models, and technology". One respondent said that the initial gain would be for "Designers and Contractors", but the final gain will be to "UK Taxpayers" through the proper delivery of BIM projects. This person also added, "I hope that BIM will reduce the cost of projects [if it is used correctly], and that this creates enough efficiency that the taxpayer benefits".

Chapter 4: Methodology

4.1 Design/Method of research

The plan defines how the empirical research should be connected to the concept. Research design defines the method of collecting data as well as the manner the results will be analyzed. The quality and type of the data collected empirically is influenced by the research method selected. I've chosen exploratory research for my research. The thesis addresses a widely-cited problem in research, namely the absence of BIM application. My study aims at understanding the barriers that hinder BIM implementation within the AEC sector. The research will also analyze the effects of various barriers that have been identified in prior research on specific projects, and identify which were considered to be the most troublesome. This research plan is a way to study barriers' impact in deciding whether to implement BIM or continue using traditional methods. The research plan is designed to develop a theoretical framework which explains the various barriers that could hinder the AEC sector from adopting BIM.

There are two different ways of finding out what's correct or wrong and making conclusions. Deduction and induction are the two ways to make conclusions. Induction is based on logic, while deduction is based on the evidence.

Induction can be used to make generalizations from the observations. Inductive research progresses by observing, finding and the construction of theories. In order to improve the theories of existing literature, new findings are included. The research results in the theory. The research is part of the category of qualitative research. It is crucial to remember that reasoning based on inductive logic cannot be completely reliable. These conclusions are based on evidence-based observations. You can't be sure that intriguing results aren't missed when there are a lot of observations.

The use of logic can be used to draw a conclusion. Researchers construct hypotheses on the basis of the existing information. They can be verified through experiments, and then accepted or denied. The type of research that is conducted tends to be a part of research that is quantitative.

The thesis incorporates the use of inductive reasoning to an extensive degree. This type of study that I am conducting will enable me to draw conclusions from the information I have collected, and further expand my knowledge of the barriers which hinder the application of BIM. In the second instance I employed the deductive method to determine the barriers that were impacting different projects. After that, I modified the interview and cases in accordance with the findings. There are two major categories of techniques for research that are qualitative and quantitative. The two research methodologies differ in how data is collected. The choice of method is based on the research's design. Research that is qualitatively based is the best choice for exploratory research. Qualitative research methods is a great way to concentrate on learning the root of the issue.

Through a review of the literature on BIM and analyzing the literature, we were able determine the most significant BIM associated challenges. EFA is also used in testing the reliability of concepts. It was used to evaluate the reliability and non-dimensionality of measurement elements. The principal component analysis method was selected due to its ease of use and reliability. Varimax was selected as the Varimax rotation was chosen over straight oblimin or Promax due to its more evenly distributed load among the variables. The factor analysis was conducted on the 100 questionnaires filled out, along with the 35 variables identified.

In order to study the challenges of BIM We used the structural equation model (SEM) to reveal the interplay between unobservable and many variables. SEM was utilized to test different models for the relationship among BIM obstacles. BIM challenges. Byrne asserts that SEM has become increasingly commonplace for non-experimental research, particularly in cases where hypothesis analysis techniques are not strictly adhered to. The model of PLS was employed to study the BIM barrier-related relationship in line with the study's goal. It included both formal as well as reflective variables. In order to evaluate the effect on the BIM barrier, the study looked at three additional major assessments including the measurement model as well as the structural model. CMB can be used to explain the reason the reasons why results from examinations are not accurate.

The inaccuracy could be explained by the process of collecting data. It is crucial to know about the issues, and to be able in determining if CMV is present. To determine if there was a CMV in the area, it is crucial to ensure that the issues and difficulties be identified. The model of measurement was selected in light of the discriminant and convergent validity. The proposed model (i.e. investigating the notion) has been successfully applied. Research design is an approach to discover the answer to the issue. conducting surveys, collecting data and conducting professional interviews in accordance with research subjects and reviewing or conducting diverse cases studies that relate to a defined topics within the conceptualization as well as analyzing those case studies to determine if the goals are achieved. Implementing these strategies could affect the information which will be used in the dissertation.

S/N	Problem
1	Lack of government, clients, and contractor support
2	Failures in technological support
3	High cost of BIM application and inadequate BIM awareness
4	The construction industry's lack of trained professionals
5	Accessibility and cost of specialized BIM software
6	Computer self-efficacy
7	Lack of information technology infrastructure to enhance BIM use
8	Challenges in implementing new forms of teamwork
9	Resistance to change of professionals in the construction industry
10	The failure to retrain professional members in the use and application of BIM
11	Problems with BIM interoperability at every stage of a project
12	Lack of BIM cooperation guidelines and standards
13	Data privacy and data ownership issues
14	Lack of managers' awareness and support
15	Contractual environment
16	Inefficient BIM education on collaboration
17	Failure to acquire individual BIM knowledge
18	Lack of reference materials to recommend BIM application to Professionals
19	Lack of qualified BIM experts
20	Not having sufficient knowledge when it's needed
21	Problem of BIM application incompatibility
22	Frequency update on software
23	Fragment nature of the construction industry
24	Lack of initiative and education
25	Conflicts between project managers, information technology managers, and building information modeling managers
26	Fear of Safety and reliability of building information modeling
27	Cost of required hardware upgrade for BIM
28	Lack of common data environment
29	Lack of standard BIM protocols for cross-industry collaboration
30	Lack of standards to guide the implementation of BIM
31	Complicated nature of BIM tools
32	Awkward team configuration and structure
33	Team members tend to work in isolation during projects
34	Opposition to information sharing
35	Designers and the supply chain downstream have not established a reliable method of working together

Figure 9: List of barriers with the serial numbers

Self-administered surveys were given to professionals in construction, such as. engineers, architects surveyors engineers, project managers and engineers that are members of the body of professionals. The number of questions that were administered totaled 261, out of which 102 were gathered for analysis. The questionnaire contained information on the highest degree of qualifications, experiences and the the number of projects they are currently working on. The questionnaire also inquired about whether they were members of an organization that is professional and how they value initial estimates. The data contained in these documents were extremely helpful when discussing the results.

It's the measurement model which defines what the world looks at the moment with respect to a latent component. To evaluate the BIM barriers in the PLS/SEM process, the validity of discriminant as well as convergent is to be assessed. Convergent validity is measured in conjunction with construct validity. This is the degree to which the barriers identical have the same logical structure and are consistent.

A composite score of reliability (rc) as well as Cronbach's alpha(a) as well as the average variance taken (AVE) are able to determine the convergent validity of the suggested concepts within PLS-SEM. Table 1 indicates that every one of the BIM barriers are characterized by scores of composite reliability that are over the acceptable minimum of 0.60. The Cronbach Alpha was also above the minimum acceptable value of 0.60 and demonstrated a moderate to moderate level of reliability, as suggested by Perry and. and. Similar AVE tests were utilized to assess the validity of convergence tests for construct variables employing the equation below.

$$AVE = \frac{\sum \lambda_i^2}{\sum \lambda_i^2 + \sum var(\varepsilon_i)}$$

The mean variance is called an AVE value, where li is the load of each component element on latent variables and var(ei) (which is 1 less l 2 which is called the variable. The value of AVE calculated by PLS 3.0 exceeded 0 which indicates that the measurements are internally stable and converging. The results are shown in the Table. But, Hulland says that in the context of an explanation, 0.40 or more is sufficient.

4.2 Methodology used

The methodology used in these models is to analysis, study and review of case studies previously conducted. This will help to identify the core of the objectives and the potential BIM impacts on AEC. My research aims to demonstrate that the potential of BIM isn't restricted to design or the construction stage and also has value during the operational stage of buildings as well as their lifespan.

This thesis will examine using case study, the way in which BIM was used at various phases of construction. The thesis will also analyze the barriers and advantages that are still a problem to be overcome in BIM implementation. The thesis employs an approach to research that involves studying case studies in order to gain a comprehension of how the latest technology will help the stakeholders control the flow of information throughout the construction life cycle. It will be more efficient and reliable over conventional approaches; however, it will not provide a description of the conventional techniques. The thesis will review case study that demonstrate the ways in which BIM implementation at earlier phases as opposed to later in the operational phase, can enhance the management of buildings and improve their efficiency.

4.3 Qualitative method

The study utilized the qualitative method of research since it matched the research's context and characteristics. As per Taylor and coworkers, "qualitative methodology refers in the broadest sense to studies that produce qualitative data, such as the individuals' spoken or written words as well as their behavior that is observable". To conduct the research, a variety of executives at various levels were surveyed. The method was selected due to the nature of the research. Additionally qualitative research goes beyond established theories, models and hypotheses, to the development of new ideas as well as understandings.

The case study is the majority of it comes in the process of production that is being studied. The information gathered from the actual case study can be utilized to improve academic research, and to examine it against what's actual experience in the project. The case study is finished by the designing stage. The project is at the beginning of production. The case was chosen by the authors due to its distinctive feature. Veidekke Max, the prefabricated foundation, and the other components provide another motive for selecting this specific case. This is the first time for the firm to integrate BIM with this function. They are keen to work on the project.

4.4 Results approach

There are two methods to conduct your research, or to write an academic paper. Qualitative or quantitative research techniques can reveal what the distinctions are between them. Methods used for collecting information for a thesis determines the distinction between these two methods.

4.5 Data collection

The research questions form the foundation of the research. All the research, data and information collected as well as the information gathered directly responds to these inquiries. The research's origins as well as the methods used to conduct research should also align. To achieve this objective, observational and direct interviews are employed. The form and nature of information collected is defined by the research's purpose and change over time.

The research methodology chosen is realist research. There are a variety of research studies that have been exploratory and descriptive of Building Information Modelling. It is crucial to conduct a research study about the ways in which BIM could be used for construction projects in the Middle East, KSA and the region around. This can help determine the issues that remain within the construction industry. Realists believe that the universe is real, and this research will explore the reality of it, and will include the implementation of BIM across all of the Middle East and Saudi Arabia.

Therefore, it is essential to collect data specifically relevant for Saudi Arabia and the Middle Eastern construction market and Saudi Arabia and to comprehend the nature of problems and how intense that they have within the construction industry. To comprehend the practical applications of BIM We will gather and analyse qualitative data utilizing the philosophy of research that is based on real-world thinking. Research goals are discussed. Interpretivism allows for the use of qualitative information utilized to conduct study. The research conducted is to study the real-world issues, and also to suggest solutions to the KSA construction sector. It is therefore crucial to obtain reliable and valid results. The results will be evaluated in accordance with the views and opinions of the research participants.

4.6 Research strategy

Many research methods are being developed to aid in exploratory, descriptive and explanation research. They are classified into two groups that are inductive and deductive. Each research method has distinct advantages and disadvantages. Strategies for research are selected according to a range of elements, such as the research questions that need to be addressed and research objectives, the duration of study and knowledge levels, the level of research methods, the philosophical foundation as well as the resources available. They do not provide the exact solution to a research question nor are them in any way mutually distinct. To determine the most effective research method It is essential to consider the pros and cons of each.

Chapter 5

5.1 Results and Analysis

BIM adoption in Nigeria and other developing countries is slower than it is expected to be compared with more advanced economies. BIM adoption is hindered in developing nations by a number of factors. These include a shortage of support from government officials and contractors, inadequate training of professionals in BIM usage and applications, lack of education and initiative, as well as an inability of changing existing working practices. BIM adoption in Nigeria has been slow, both within the public sector as well as amongst construction companies. BIM is often used by architects to improve the presentation of their work. In Nigeria, the use of BIM is restricted by the various specialized groups attracted to this new technology. This is because these bodies have not kept up with technological advancements. BIM adoption is hindered by a number of factors, including fear of change, the high costs of BIM implementation, the lack of BIM skilled labor, the general lack of client interest, the scarcity of BIM trained workers in the industry and the absence of BIM qualified personnel.

The primary focus of this aspect is technology, which includes application and software compatibility; quality monitoring and authorizations, visualization and detection of clashes in layouts, as well as BIM protocols and standards. It is intended to be the foundation of the BIM Interface. Researchers summarize existing research and propose future study topics to counteract this risk. A lack of integration can lead to an inaccurate impression about BIM among non-BIM or non-construction professionals. BIM is not well understood by non-construction professionals and BIM practitioners alike. There are no comprehensive lists of BIM advantages and cost savings. BIM software is expensive in Nigeria. The majority of construction firms cannot afford to purchase computers, and all the expensive extras that go with them. This includes software. It is expensive to purchase the software and install it on every computer system, so the older versions of BIM are used. The cost of learning BIM is also high. The widespread perception that BIM assessment is difficult to implement is a major obstacle.

The table below shows the sum of rotations for loads that are squared include 67.573. It is more than 50%, and suggests the ability to utilize EFA. This table has a significant impact on each obstacle based on the varimax of rotation. The names given to the elements listed in the Table are not determined by a particular method however, they were justifications based on the background of the researcher and expertise.

Constructs	Barriers	Loading
BIM literacy among the construction professionals	B9	0.680
	B10	0.750
	B4	0.520
	B24	0.554
	B23	0.687
	B17	0.556
	B18	0.564
	B11	0.850
	B7	0.780
	B35	0.654
	B8	0.950
BIM collaboration and standard	B16	0.856
	B34	0.687
	B12	0.786
	B32	0.569
	B22	0.785
	B31	0.654
	B30	0.458

Figure 10: BIM literacy among the construction professional

	B5	0.965
	B2	0.650
Cost Impact of BIM	B3	0.856
	B6	0.654
	B28	0.576
Problem of standardization	B29	0.789
	B33	0.657
	B25	0.756
Competitive mentality among the stakeholders and BIM Reliability	B26	0.650
	B27	0.860
	B15	0.650
Contract condition	B14	0.756
	B13	0.745

Figure 11: Cost impact of BIM

The variance due to bias of common methods is utilized to determine the variance of errors and to determine the validity of an analyses. The model was then subjected to an analysis of a single factor to determine variations that is caused by the traditional method. The general consensus is that bias in the standard method does not affect the results when the variance falls below 50 percent. The current study, it was discovered that the variations in the procedure are not impacting the data since only the first component is responsible for 42.23 percent of the overall variation.

Most educational institutions lack IT experts and teachers who are comfortable using BIM in the classroom. BIM is not widely used in the Nigerian construction industry due to a lack of information technology infrastructure. Internet and computer accessibility are among the poor IT infrastructure of universities. It is unlikely that faculty and students have the intelligence or information management skills to fully exploit all the opportunities available. Since they are widely recognized as being instrumental in the classroom, teachers can greatly benefit from accessing high-quality resources.

Textbooks can help first-year teachers feel confident and secure. Teachers do not provide textbooks and other BIM reference materials to students. There are not enough BIM experts, which hinders the spread of BIM paradigms and marketable BIM information. Mehran says that the organization's structure and dimension affect BIM adoption. This includes the availability of BIM specialists, training and certification in BIM technology, as well as support from clients and top management.

The results were in agreement with Ugliotti. He said that mismatched procedures, technology, personnel and processes were just two of many obstacles to BIM adoption. These problems are encountered throughout all phases of operations and maintenance. Saka and Chan determined that the biggest obstacles to BIM adoption in Africa were people and process limitations and barriers. BIM must be integrated into the regular operations of a company. The government has to provide assistance and develop a communication method that will increase the use of BIM by organizations. They examined in this study the major obstacles that prevent the use of BIM by the Iraqi Construction Sector. Their research found that the main obstacles to BIM adoption are the lack of BIM investments, the shortage of professionals, an absence of national BIM standards, as well as the unwillingness to adopt change.

BIM, as mentioned earlier, should be used for describing the goal of the design and including the designer's previous knowledge. In order to encourage experts to become familiar with BIM, it is important to continue investing in BIM-related research. Many construction firms in developed countries are using BIM techniques to improve the traditional approach. This has led them to achieve greater success. According to the research, and to the literature reviewed, there are a number of challenges that have contributed to the lack of BIM implementation by Nigerian construction experts. These include a lack of computer proficiency, a lack of BIM standards for collaboration across industries, resistance of construction professionals to change, and a lack of infrastructure to support BIM. The lack of a working relationship between designers, including architects and civil engineers (to name a few), and the downstream supply chains was the biggest challenge, preventing the adoption of BIM Technology by the industry.

This is due to the discrimination that exists between construction professionals. These challenges are also a result of the education system in the country, which does not encourage smooth collaboration among construction professionals. Due to the challenges faced by construction companies in Nigeria, BIM has not been fully maximized. To improve this industry, certain strategies will need to be adopted and adapted. A lack of computer proficiency among professionals is also a result of the nation's economy, as BIM workstations are expensive and require hefty graphics. The survey revealed that lack of BIM standards for collaboration across sectors and resistance to change by construction experts are the two most important factors in preventing widespread adoption of BIM. The 35 challenges to BIM implementation in the construction industry were divided into three categories: opposition to sharing information, fears of reliability and safety of BIM, and failures of technology support. The participants have already seen how BIM technology has overcome these challenges. BIM obstacles are evident in the discussion above. They impede BIM adoption and development. Several engineering projects suffered significant setbacks as a result. In previous studies, a combination of a literature review and a survey questionnaire was used to investigate the BIM challenges. Researchers have done a great job in identifying the specific obstacles, but have paid less attention to the relationships between these barriers and their effects on each other. These studies employ realistic research methods and a fresh perspective in order to analyze the BIM challenges.

Dong's study is impressive. Zhou and colleagues identified six barriers to implementing the BIM in China. These include a lack government leadership, organization challenges, legal concerns, high costs of application, the challenge of changing one's thinking, and the lack of external incentive. Ozorhon & Karahan also investigate the factors that influence building information modeling adoption in developing nations where BIM has just begun. Ma et al used the same method (principal components analysis) to examine the reasons for the low BIM usage in AEC in China. In the main component analyses, the factors identified were: expertise and abilities, technical conditions and system inertia. Additional inputs, work changes, adoption risks, as well as extra input and the inertia of the system.

Productivity is a problem for the AEC industry. The productivity growth in the AEC-industry is much lower than other industries. This problem, as described in the chapter on theory, is thought to arise from the fragmentation of the AEC industry, coupled with the need for collaboration. Collaboration is needed to ensure that a project can be completed by many actors. BIM addresses these problems by improving the exchange of information between the team members and during the life cycle of the building. There is no agreement on the current definition of BIM, as described in the chapter about theory. There are some elements that are shared, but the definitions vary greatly in their approach. The majority of definitions refer to BIM in terms of 3D models, a form that is rich with information and uses objects. It is not always clear how this technology will affect the processes of the industry.

5.2 Totally latest concept

BIM is actually a way to transfer information about the building. The information can be exchanged between people within an actor, or even between actors. This process is available from the beginning of the project to the end. The transfer of more information efficiently will not improve projects, but it can enable better work practices. As Kiviniemi says, BIM as a whole isn't a goal but a tool for achieving other goals. The tool can be applied in various ways depending on the goal that each actor wants to achieve. BIM does not exist as a standalone goal, so it's important to understand what the actual objectives of BIM adoption are. The goals of the BIM adoption will determine how BIM is used and what changes are needed within the team.

It is not necessary for the work of other participants to be affected if, say, a particular actor wants to use a BIM to enhance their performance. Coor, the facilities manager in the NKS project, is a good example of this. Coor is keen to use the model extensively, but the work processes of other parties do not need to be changed much as long as it's available. If on the contrary, the goal is to increase the overall productivity of the project, then more changes are needed in all work processes. BIM allows for a change to the fragmented or inefficient work processes in the AEC industry.

The way to change these inefficiencies is not through the use of ICT, but by improving the processes that depend on them. Poor collaboration and fragmentation is a very complex issue. Implementing a more effective information sharing system won't change everything overnight. BIM can be used to solve the problem of a fragmented and uncoordinated industry by creating new processes. The new processes will improve the collaboration, and therefore increase productivity. The improved productivity does not depend on a single person, rather it requires that all participants in the project adapt to new processes. BIM has been cited by many researchers as having economic and other benefits. However, there is also an interesting causality aspect. BIM can be used to enhance processes, but the benefits are derived from improved processes. It is therefore important to know the goals of the project, and then adapt BIM's use.

5.3 BIM is all new technology

Researchers have stressed the importance of interoperability and problems that arise when software providers do not work together. Steel et al state that the AECindustry is collaborative and requires an efficient information exchange between all parties involved. On the other hand, BIM tool development has been driven to more proprietary formats that are connected with specific tools. It is a major barrier to BIM implementation because the technology encourages collaboration. A great deal of value can be lost if it's not possible to combine different models due to lack of interoperability. As a solution to the problem of lack interoperability, open formats were presented. Steel et al present IFC as a standard that is open and has many options, however this does not achieve total interoperability. IFC's interoperability will suffice in most situations, however, an open standard may not be as effective as proprietary options.

When the most interoperability possible is required, it may be better to use a single family of software. Discussions between consultants led to the decision of which formats would be used. So long as formats that a consultant wants to use are compatible with other consultants, any format can be used. Skanska, on the other side, wanted to make sure that all BIM model deliveries were made using a certain format. Interoperability was achieved in the NKS Project, avoiding the problem of open formats. Simultaneously, the BIM model in this project was primarily produced to meet the contractual requirements set forth by Stockholm County Council.

It is difficult to determine if the BIM solution offered by Skanska actually improved collaboration among actors, or merely provided a platform where everyone could combine their own models with those of Skanska. When models are used extensively by actors other than their creators, interoperability becomes more critical. Steel et al describe that only limited interoperability levels are required to combine the models of different actors. It is possible that the level of interoperability achieved by the NKS method will not suffice to allow collaborative processes to be implemented in the project to increase productivity. Interoperability is a major obstacle to BIM. However, as Kiviniemi explains working technology can only be an enabler. Interoperability solutions haven't been used to drive BIM adoption in the studied cases. Even better interoperability is unlikely to have led to an increase in the use of models by itself. When the level of interoperability does not allow for the desired change in the work practices, this will inhibit BIM adoption.

Chapter 6: Conclusion

The best part of the project is to fulfill all the aims and objective section such as relevance of BIM with the present construction and process. The project has showed the exact pattern of working of BIM through the case studies also which has played a vital role to make the project understandable.

In order to maximize profit without losing functionality of the construction project, it is important that successful concepts are used throughout all stages of project lifecycle. Many studies focused on BIM drivers, but very few examined the impacts of BIM on the construction development. The purpose of this research is to identify and solve the challenges associated with BIM implementation in construction for developing countries. A thorough literature review was performed to identify BIM-related obstacles. Then, an exploratory factor (EFA), was conducted to categorize these challenges. A questionnaire was used to survey 100 Nigerian construction professionals. The results were then used in the partial least-square structural equation model (PLS SEM). Model conclusions identified the biggest implementation obstacles for BIM. These conclusions can be used as a guide or guidance for developing nation policymakers who want to complete projects by using BIM and avoiding BIM obstacles.

The model generated by this research investigates BIM's major challenges. Policymakers such as regulators and government agencies can use these challenges to create a strategy that will increase BIM usage in the AECO industry. In the first part of the study, the researchers assessed the biggest challenges in implementing BIM within the construction industry. The study lays the foundation for future research on the BIM implementation challenges in the AECO industry. This research has also made several theoretical and practical advancements, such as:

This study is a contribution to theory by providing new concepts that could be incorporated into the framework. The challenges of BIM implementation can affect BIM understanding and adoption at every stage in a project.

The research in Nigeria on BIM is in its early stages. The research addresses this need by concentrating on the issues that directly relate to BIM implementation and those that hinder it. This study is the first to use a predictive model that can assess how BIM implementation obstacles will impact BIM awareness and usage across all project phases in the AECO sector. This resource should help accelerate BIM's spread in developing nations. The contribution is empirical, as it attempts to do what has never been done before: evaluate the theoretical links between two variables (BIM implementation obstacles and BIM awareness and usage in project lifecycle).

The study involved a collaborative global experiment between WSU and NICMAR Pune in India. Two aspects of the study were examined from a sociotechnical point of view: The role of the project team in temporary organization and the BEP strategy as an intervention pre-process. Findings showed virtual coordination of project tasks to be a challenge. To cope, organizations adopting similar business models must consider strategies for intervention during or before the process. BIM Execution Plan enabled members to define their roles and anticipate project complexity. It also allowed them to plan ahead a priori. The participants noted that the BEP needed to be planned and implemented more efficiently, even though not all members adhered to the plan. Individual and team observations included a proactive approach, and an awareness of institutional differences between nations. A socio-technical perspective provided interesting insights. Due to the short duration of this study, and its limited amount of data, it is not possible to generalize or explore further.

For a more holistic understanding it is recommended to explore the use of similar studies with longer duration projects to determine whether pre-process, process, or post-process intervention can be used to promote successful outcomes for global virtual collaborations. Kunz and Fischer note that large companies will soon be using Virtual Design and Construction in their geographically and technologically diverse operations, which could lead to the proliferation of global sourcing. The study serves as a step in that direction. This exercise, in the contexts of curriculum and academics, was very well received by students at both participating universities. They described it as "rich experience". This exercise provided insight for faculty coordinators, as well as a strategy to increase accountability by defining and allocating team roles.

The industry also needs to have time to understand and develop the tools to reap the benefits of BIM at all levels, and during all phases of a project. It is necessary to develop a way of measuring the impact BIM has. The client with the greatest interest in the project is the one who should demand BIM. To see benefits, one needs to work on two BIM projects. BIM projects are still relatively new, so the industry will need some time to get used to them. It is important that the industry improves its ability to manage discipline. They can see that the understanding is growing in only these two projects, which opens up new opportunities. The FM is slowly figuring out how to use BIM efficiently for the maintenance of these complex and large projects.

Future scope

BIM is the next step in construction because of its undeniable impact on improving the effectiveness and efficiency of coordination, management as well as construction. BIM isn't just useful for architecture, but it has also proved to be efficient across other AEC sectors too. Projects that use BIM have experienced improvements in productivity ranging between 75 and 240% in accordance with research. Companies that use BIM record a 7 percent reduction in the project's duration.

Smart BIM models can be a fantastic alternative to traditional 3D concept models. They are able to provide further information such as the amount of duration required for every phase of the project, estimated expenses, as well as the the energy usage of buildings when it's built.

BIM is an effective instrument that facilitates better coordination between engineers, architects and contractors. In addition, construction projects are received a lot of focus. Despite their significance issues like facilities management and deconstruction aren't considered when conducting research. Volk et al. They also mentioned the fact that standards such as Cobie and IFC are key to the future of BIM for both new and existing structures. The integration of modern technologies with BIM BIM could also provide an efficient bridge to BIM integration in older buildings. BIM can be a useful instrument for construction. BIM is a digital intelligent model which helps control the various stages of construction projects. The research on the subject has changed from an abstract method to a more specific one. BIM offers many advantages in the construction industry, however it has limitations that prevent its development. To make BIM more efficient, it requires innovative technologies to change the industry of construction to an evolving one.

References

1. BIM is not only a way to improve productivity in a project but also an effective tool for improving performance during and after the project. Ding et. al. state that the adoption of this technology is most advantageous in three areas: digitalization, purc [Report] / auth. Oti AH. - [s.l.] : https://www.sciencedirect.com/science/article/abs/pii/S09265 80516301972, 2021.

2. Aranda Mena et al. cite incompatibility between BIM software as one of the major challenges for BIM adoption. Ku and Taiebat claim that different BIM programs are incompatible, so data from one must be transferred to another, rather than being shared. Thi [Report] / auth. Liu P. - [s.l.] : https://www.hindawi.com/journals/ace/2019/9482350/, 2020.

3. Steel et al concluded that IFC had achieved relative interoperability at the level of files and visualisations within a small subset domains. It is particularly notable in the architectural design domain. It still has challenges in situations requiring se [Report] / auth. Xu Z. - [s.l.] : https://www.mdpi.com/2075-5309/12/8/1158, 2019.

4. Nya Karolinska Solna - Construction of the hospital "Nya Karolinska". This project has high expectations for the implementation of BIM, and these will be maintained throughout the lifecycle of the project. This project is a good example to use when studyi [Report] / auth. Ren R. - [s.l.] : https://www.hindawi.com/journals/ace/2021/8824613/, 2021.

5. BIM can provide tenants with new services in addition to developing and enhancing the functionality of the NKS facilities management system. The BIM model with its linked databases allows Coor to offer many new services. There are

discussions on how the p [Report] / auth. Xu Li. - [s.l.] :
https://www.hindawi.com/journals/ace/2022/1901201/, 2023.

6. BIM is a tool which allows the storage and re-use of domain
   and information knowledge during the entire lifecycle of a
   project. BIM's main function is to coordinate and integrate the
   sharing of knowledge and information between disciplines
   within a projec [Report] / auth. NBS. - [s.l.] :
   https://www.thenbs.com/knowledge/what-is-building-
   information-modelling-bim, 2020.

7. A number of initiatives have also been put on hold due to a
   limited scale of investment. As a group, the construction
   industry of developing countries is not meeting the needs of
   governments, their clients and the society. It lags behind the
   other industr [Report] / auth. NIOH. - [s.l.] :
   https://www.ncbi.nlm.nih.gov/pmc/articles/PMC7124044/,
   2020.

8. BIM has become increasingly popular with construction
   experts around the globe. According to the National Building
   Specification, the United Kingdom (UK), Canada and Finland
   are among the advanced BIM countries. Building information
   modeling (BIM), both i [Report] / auth. ADB. - [s.l.] :
   https://www.adb.org/sites/default/files/publication/29823/infr
   astructure-supporting-inclusive-growth.pdf, 2022.

9. Nieto-Julian and colleagues claim that BIM can help members
   of interdisciplinar cultural teams by making information
   exchange easier. Stransky & Dlask have shown that BIM can
   improve project performance, and help with decision making
   throughout the projec [Report] / auth. Agenda. - [s.l.] :

https://www3.weforum.org/docs/WEF_Shaping_the_Future_
of_Construction_full_report__.pdf, 2019.

10. There is currently no clear definition of what BIM is. BIM is defined in many ways by professionals, which makes it difficult to discuss. There is no consensus among the many organisations that have attempted to define BIM. Many aspects of the model are t [Report] / auth. UNESCO. - [s.l.] : https://unesdoc.unesco.org/ark:/48223/pf0000189753, 2023.

11. Chan and other studies have shown that the lack of skilled workers has been a significant roadblock in BIM adoption. Aranda Mena and colleagues claim that when there aren't any workers who can advocate BIM adoption, it is easy to discuss its adoption beca [Report] / auth. Marsh. - [s.l.] : Chan and other studies have shown that the lack of skilled workers has been a significant roadblock in BIM adoption. Aranda Mena and colleagues claim that when there aren't any workers who can advocate BIM adoption, it is easy to discuss its adoption beca, 2019.

12. To ensure the success of BIM, all stakeholders must work together to update, insert or modify the information within the BIM model at the various stages during the lifecycle of the facility. The model is designed to reflect and support the different roles [Report] / auth. LB. - [s.l.] : https://www.letsbuild.com/blog/recognising-bim-roles-project-cycle, 2022.

13. Tse et. al. suggest that the main reason why architects do not adopt BIM is because of the lack demand by clients and project members. In their survey, a large majority of respondents agreed that existing entity-based systems can meet their design and dra [Report] / auth. Tse. - [s.l.] : https://napier-

14. repository.worktribe.com/preview/2871440/JOBE%20HK%2 0B_B.pdf, 2019.

15. The law in Sweden prohibits the universities from owning their buildings. Instead, they are required to rent them. The facilities on the KTH Campus at Valhallavagen, are currently rented by Akademiska Hus. This is a Swedish state-owned company that owns t [Report] / auth. John. - [s.l.] : https://news.ki.se/changed-regulations-for-universities-regarding-housing-to-students, 00.

16. Nya Karolinska Solna is a complex and large construction project that began in the summer of 2010 and continues until fall 2017. The hospital will have a total area of approximately 320 000 sqm. This hospital will be built and managed by a PPP that will l [Report] / auth. DP. - [s.l.] : http://kth.diva-portal.org/smash/record.jsf?pid=diva2:951869, 2018.

17. Ahola, T., 2021. Xu et al found that the presence of a local partner is crucial for the development of an international AEC firm. Based on this research the strategic partnership that include local construction firms or design companies are the most effective collaboratio, s.l.: https://www.sciencedirect.com/science/article/pii/S02637863 21000739.

18. Pheng, L., 2019. Ling et al conducted a study to determine the most important factors that help AEC firms secure contracts with China. The capacity of the firm to understand the requirements and demands of its clients was identified as the most important aspect. This make, s.l.: https://www.ncbi.nlm.nih.gov/pmc/articles/PMC7124044/.

19. Li, L., 2022. The issue has been extensively discussed how important it is to have a well-organized construction design

management system for the success of a project. Design management has been plagued by ineffective communication, poor documentation, inadequate or mi, s.l.: https://link.springer.com/article/10.1007/s10796-022-10308-y.

20. Bender-Salazar, R., 2023. Iterative design is the process of designers finding issues, sharing information and thoughts with one another and then implementing their ideas and resolving the issue. In order to improve the efficiency of design, it is essential to improve the design p, s.l.: https://innovation-entrepreneurship.springeropen.com/articles/10.1186/s13731-023-00291-2.

21. Akponeware, A., 2018. The design of the construction is a continuous process that has numerous interdependencies. This requires the participation of all participants as well as the coordination of all project participants. It's a fundamentally collaboration process, which reli, s.l.: https://www.mdpi.com/2075-5309/7/3/75.

22. Knotten, V., 2020. Collaboration and communication when designing and carrying out a design can be seen in the variety of methods that focus on processes and are used to improve and enhance the management of construction design. Senescu et al. introduced the Design Process , s.l.: https://www.sciencedirect.com/science/article/pii/S2212567115001586/pdf?md5=b37abe5af33a0581269f1b6183003a87&pid=1-s2.0-S2212567115001586-main.pdf&_valck=1.

23. Chun, J., 2018. Design Interface Management System(diMs) is a method that manages the iterative process of design for large-scale projects is currently being developed. The Dependency Structure Matrix (DSM) can be utilized to

develop design plans and also to illustrate i, s.l.: https://www.jstage.jst.go.jp/article/jaabe/17/1/17_95/_pdf.

24. Matusova, D., 2018. Building Information Modelling, a process that is developing and able to efficiently store and manage the entire project's data (physical properties as well as functional attributes) in a single database. BIM tools are utilized for design purposes to carr, s.l.: https://link.springer.com/chapter/10.1007/978-3-319-70055-7_34.

25. Herr, C., 2019. It is worth noting that the Chinese government, in contrast to other nations, is yet to make any regulations for national use that require BIM implementation. This means that BIM acceptance in China in the past decade is heavily influenced by the market. , s.l.: https://www.sciencedirect.com/science/article/pii/S22884300 18300708.

26. Vass, S., 2019. Yung et al. reported the results of a BIM research study based on BIM of BIM-based design in China. Utilizing BIM will not necessarily reduce duration of design due to the fact that 2D designs remain heavily in. This is mainly because of the need to provi, s.l.: https://www.diva-portal.org/smash/get/diva2:874077/FULLTEXT01.pdf.

27. Jiang, Y., 2021. Eastman et al. stated that a Building Information Model was constructed by intelligent digital assembly with embedded knowledge about parametric attributes and characteristics. Kymmell has described BIM as an intelligent digital model that is connected to, s.l.: https://www.ncbi.nlm.nih.gov/pmc/articles/PMC8691975/.

28. NAIM, A. A., 2020. True BIM tools have an integrated database, where data represents the model of the project and

reports are queries or views. All plan views, sections views, elevations and callouts; as well as perspectives, are all live in a project model. Project teams a, s.l.: https://core.ac.uk/download/pdf/196185612.pdf.

29. Pan, X., 2023. BIM can be used to produce schematics and design details, improve the presentation of a project to the client for better decisions and a more clear understanding of what work is to be done. Autodesk said that BIM was used to analyze energy efficiency and , s.l.: https://www.sciencedirect.com/science/article/pii/S20904479 23001417.

30. Sahil, 2020. The challenges of BIM adoption and implementation at the organisational level are related to policy changes, changing working cultures, costs, and people. Fear of change is the biggest barrier for BIM implementation and adoption. When people succeed at so, s.l.: https://mountainscholar.org/bitstream/handle/10217/173492/Sahil_colostate_0053N_13498.pdf.

31. Zaia, Y., 2023. London et. al. highlighted that the adoption of BIM at the organisational level is determined by the relation between the current working practices and the future BIM scenarios as perceived in the organisation's policies. If an organization is at level ze, s.l.: https://www.sciencedirect.com/science/article/pii/S20904479 22001927.

32. PiC, 2020. In the construction industry, productivity is of great importance. It is the best indicator for how much money and time will be spent per task and on the final project. Chelson quoted Warren as saying that "productivity in construction means completing co, s.l.: https://web.mit.edu/parmstr/Public/NRCan/nrcc37001.pdf.

33. Wrike, 2020. Contractors have traditionally used various project management tools, including WBS (Work Breakdown Structure), to plan construction activities. WBS (Work Break Down Structure) is a hierarchical, deliverable-oriented decomposition that breaks down the tas, s.l.: https://www.wrike.com/project-management-guide/faq/what-is-work-breakdown-structure-in-project-management/.

34. NAIM, A. A., 2020. A good document management system is another important factor in contractor productivity. It allows the latest and correct information to be sent to all stakeholders at any given time. These systems include Meridian, Primavera and Bentley, as well as proj, s.l.: https://core.ac.uk/download/pdf/196185612.pdf.

35. TIDF, 2019. Flager and coworkers found that designers can be spending as much as 58% of their time during the design phase coordinating information and coordination. This involves manually integrating specific representations of the design as well as analytic models., s.l.: https://www.interaction-design.org/literature/article/5-stages-in-the-design-thinking-process.

36. Li, S., 2019. Design Interface Management System(diMs) is a method that manages the iterative process of design for large-scale projects is currently being developed. The Dependency Structure Matrix (DSM) can be utilized to develop design plans and also to illustrate i, s.l.: https://www.researchgate.net/publication/245136298_A_web - based_system_for_design_interface_management_of_constru ction_projects.

37. Zhang, J., 2020. BIM is, on the other side is a management of information tool that simulates the construction and design

process instead of merely a visual representation. BIM surpasses 3D modeling and allows designers to choose the best design for their needs by modelin, s.l.: https://www.sciencedirect.com/science/article/pii/S20904479 2200421X.

38. Abioye, S., 2020. BIM software is a complex tool that requires specialized developers and vendors. Users in the construction industry must rely on these companies to provide BIM applications and tools for varying uses. Software vendors have been driving development of BIM , s.l.: https://www.sciencedirect.com/science/article/pii/S23527102 21011578.

39. IDF, 2022. Iterative design is the process of designers finding issues, sharing information and thoughts with one another and then implementing their ideas and resolving the issue. In order to improve the efficiency of design, it is essential to improve the design p, s.l.: https://www.interaction-design.org/literature/article/design-iteration-brings-powerful-results-so-do-it-again-designer.

40. Ibn-Mohammed, T., 2021. Modern management of design in China is a relatively new concept. But, given China's rapid expansion of its technologically advanced society and its economic growth over the next few years it is likely that we will witness rapid developments. In China the, s.l.: https://www.ncbi.nlm.nih.gov/pmc/articles/PMC7505605/.

41. Satyanaga, A., 2023. Succar describes BIM as an interconnected set of policies, processes and technologies that produce a new method to manage design, construction, and operation of construction projects throughout their entire lifecycle. Gu & London claim that BIM is a digit, s.l.: https://www.mdpi.com/2412-3811/8/6/103.

42. NAIM, A. A., 2019. According to an analysis of BIM, software companies that produce the software and technology for BIM implementation are in favor of the view "BIM is a technology". The "BIM as a process" or "activity" viewpoint is supported by the professional institutes , s.l.: https://core.ac.uk/download/pdf/196185612.pdf.

43. Zabin, A., 2022 . The overall process of CAD is therefore unstructured and error-prone. It's also non-collaborative, inefficient, and not collaborative. BIM workflows, on the other hand, are built upon the idea of collaboration and integration to create a centralized repos, s.l.: https://www.sciencedirect.com/science/article/abs/pii/S14740 3462100224X.

44. Lin, Y., 2020. Hardin, Eastman et al. agree that BIM can be used to analyze the operation of a building within a BIM-enabled environment. Anumba and colleagues state that BIM extends beyond the construction and design phases to facility management, with the goal of main, s.l.: https://www.hindawi.com/journals/jat/2020/8834389/.

45. Kjartansdóttir, I., 2020. BIM offers a range of new software applications and tools that can be used for creating, analysing and managing BIM models at all stages of the project lifecycle. BIM tools store information in BIM models databases, which remains there until it is modifie, s.l.: https://www.ciob.org/sites/default/files/M21%20%20BUILD ING%20INFORMATION%20MODELLING%20-%20BIM.pdf.

46. Salleh, H., 2023. BIM's benefits and productivity gains are widely acknowledged. These are becoming increasingly apparent as technology and process adoption matures. BIM implementation and adoption on actual projects are not

without their challenges, which is slowing down , s.l.: https://www.mdpi.com/2075-5309/13/6/1469.

47. Kineber, A., 2023 . BIM adoption is hampered by information bottlenecks and lack of content in project vendor products. There are also challenges with BIM application and understanding. Froese also stated that BIM's impact on project management processes and practices cannot, s.l.: https://www.mdpi.com/2076-3417/13/6/3426.

48. Shojaei, R., 2021 . The final challenge is the implementation of BIM at product level. These challenges are technical in nature and relate to BIM development and its application within projects. BIM is an entirely new technology that has triggered the implementation of BIM i, s.l.: https://www.sciencedirect.com/science/article/pii/S00401625 22005571.

49. Kunkatla, C., 2022. In the construction sector, productivity and production are terms that are used often interchangeably. However, there are significant differences between them. Production is defined as work that is done to achieve a goal or task, whereas productivity is a, s.l.: https://www.sciencedirect.com/science/article/abs/pii/S22147 85321078494.

50. RIBA, 2020. In a typical construction project, the architect is responsible for designing and producing construction documents while engineers take care of technical details. This happens in separate disciplines. Lead architects are often responsible for the coordina, s.l.: https://www.architecture.com/-/media/GatherContent/Test-resources-page/Additional-Documents/2020RIBAPlanofWorkoverviewpdf.pdf.

51. Putu, A., 2019. Contractors also have a responsibility to assess the constructability of designs and determine how they can be built using the most effective methods. Chelson reported that improving constructability tends to reduce the cost of construction by 6%-10%. Lat, s.l.: https://www.matec-conferences.org/articles/matecconf/pdf/2019/25/matecconf_i cancee2019_02023.pdf.

52. Demirkesen, S., 2021. The lean construction technique emerged to focus on activities that add value and reduce or eliminate unproductive work. Lean construction principles were borrowed by the automobile industry. Toyota Production System is a system that promotes continuous a, s.l.: https://www.intechopen.com/chapters/75657.

53. Hong, Y., 2019. To gain an advantage, construction executives and project managers need to be aware of the costs of BIM implementation as well as the expected savings. BIM value is calculated by subtracting the costs of BIM from the money saved through increased producti, s.l.: https://www.itcon.org/papers/2019_33-ITcon-Hong.pdf.

www.ingramcontent.com/pod-product-compliance
Lightning Source LLC
Chambersburg PA
CBHW050446290526
45786CB00006B/2180